Stephen Donald Huff, PhD

GRAPH MODELS FOR DEEP LEARNING

An Executive Review of Hot Technology

CAPITAL IDEATIONS LLC

Published by Capital Ideations LLC
2733 Palermo Ct.
League City, TX 77573

© Monday, August 20, 2018
First Edition Published 2018

Copyright © Stephen Donald Huff, 2018
First published in the United States of America by Capital Ideations LLC, 2018

All rights reserved. No part of this publication may be reproduced, stored in a retrieval system or transmitted, in any form or by any means, electronic, mechanical, photocopying, recording or otherwise, without prior permission of Stephen Donald Huff, PhD (Capital Ideations LLC)

ISBN-13: 978-1723761263
ISBN-10: 1723761265

For human intelligence – long may it reign supreme.

Table of Contents

FOREWORD 1
    COURSE OVERVIEW 3
        Course Prerequisites 3
        Acquirable Primary Skills 4
        About This Executive Review 4

COURSE INTRODUCTION 7
    ABOUT THIS INTRODUCTION 9
        Relevant Nomenclature 9
    STATISTICAL INFERENCE AND STATISTICAL MODELS 11
        Contrasts with Descriptive Statistics 11
    MODELING ASSUMPTIONS 13
        Three Levels of Modeling Assumptions 13
    APPLICATION-SPECIFIC METHODS 16
        Examples 16
    PREDICTIVE MODELING AND MACHINE LEARNING 20
        Predictors and Classifiers 20
        Examples 21
    GRAPH THEORY AND NEURAL NETWORKS 25
        Applications 25
    GRAPH MODELS AND DEEP LEARNING 28
        Relevant Nomenclature 28
        Applications (Graph Models) 29
        Applications (Deep Learning) 30
    LESSON CONCLUSION 32

DESCRIBING MODEL STRUCTURE WITH GRAPHS 33
    LESSON OVERVIEW 35
        Relevant Nomenclature 35
    LOGICAL REPRESENTATION OF GRAPH STRUCTURE 36
        List Structures 36
        Matrix Structures 37
    IMPLEMENTATION 39
    EXAMPLES 41

  Lesson Conclusion 43

DEEP LEARNING AND GRAPHICAL MODELS 45

  Why Deep Learning? Why now? 47
  Lesson Overview 50
   Relevant Nomenclature 51
   Essential Deep Learning Concepts 51
  Understanding Deep Learning 53
   Universal Approximation 53
   Probabilistic Inference 55
   Operational Abstraction within Deep Learning 55
  Artificial Neural Networks and Deep Neural Networks 57
   Artificial Neural Networks 57
   Deep Neural Networks 58
  Lesson Conclusion 59

MONTE CARLO METHODS 61

  Lesson Overview 63
   Relevant Nomenclature 64
  Overview (Monte Carlo Methods) 66
  Examples 68
  Lesson Conclusion 72

APPROXIMATE INFERENCE AND EXPECTATION MAXIMIZATION 73

  Lesson Overview 75
   Relevant Nomenclature 75
  Overview (Approximate Inference) 77
   Examples (Approximate Inference) 77
  Overview (Expectation Maximization) 80
   Iteration 80
   Deficiencies 81
  Expectation Maximization *versus* Variational Bayesian Methods 82
   VB Differences 82
   EM Differences 82
  Lesson Conclusion 84

DEEP GENERATIVE MODELS 85

  Lesson Overview 87

DIFFERENCES (DISCRIMINATIVE VS. GENERATIVE MODELS) 87
OVERVIEW (DISCRIMINATIVE MODELS) 88
   Examples (Discriminative Models) 89
OVERVIEW (GENERATIVE MODELS) 91
   Examples (Generative Models) 91
LESSON CONCLUSION 96

APPLICATIONS TO CHARACTER RECOGNITION, NATURAL LANGUAGE PROCESSING AND COMPUTER VISION 97

   LESSON OVERVIEW 99
   OVERVIEW (CHARACTER RECOGNITION) 100
      MNIST Database 100
      Examples (Character Recognition) 101
   OVERVIEW (NATURAL LANGUAGE PROCESSING) 102
   RELEVANT NOMENCLATURE 104
   SYNTACTICAL ANALYSIS 106
   SEMANTIC ANALYSIS 109
   DISCOURSE ANALYSIS AND SPEECH RECOGNITION 112
   OVERVIEW (COMPUTER VISION) 113
      Recognition 114
      Motion Analysis 114
      Media Recovery 115
   DEEP LEARNING AND COMPUTER VISION 117
      Overview (Convolutional Neural Networks) 117
      Convolutional Layers 118
      Pooling Layers 119
      Fully Connected Layers 119
      Receptive Field 119
      Weights 120
      Filters 121
   LESSON CONCLUSION 123

CODE BASE 125

   OVERVIEW 127
      Keras Samples 127
      Python 129
   MULTILAYER PERCEPTRON (MLP) FOR CHARACTER RECOGNITION OF MNIST DATABASE 130

Multi-Layer Perceptron for Topic Classification 133
Lstm (Long Short-Term Memory) (Recurrent Neural Network [RNN])
.... 137
Convolutional Neural Network (CNN) for Convolutional1D (Text Classification) 141
Convolutional Neural Network (CNN) for MNIST Database (Character Recognition) 145
Variational Auto-Encoder (VAE) a Generative Adversarial Network (GAN) 149

COURSE CONCLUSION 155

A Closing Analogy 157

GLOSSARY OF TERMS 159

FOREWORD

Foreword

Course Overview

This course provides a detailed executive-level review of contemporary topics in graph modeling theory with specific focus on Deep Learning theoretical concepts and practical applications. The ideal student is a technology professional with a basic working knowledge of statistical methods.

Additionally, to better inform the *interested* student, the final lesson of this course presents samples in Python describing the essential implementation of basic model structures. To reduce space and improve clarity, this code targets a basic Keras environment – this inclusion is not meant as an endorsement of one system over another (all provide benefits); instead, at the time of this writing, Keras simply offers a popular, facile 'frontend' for managing TensorFlow and Microsoft Cognitive Toolkit deep learning systems, all using this popular script.

Course Prerequisites

Although this course presents its subject matter from the perspective of an executive review, the student will benefit from modest experience with a handful of fundamental technical concepts. Specifically, the student will benefit from: a basic understanding of application-specific ('traditional' or 'rule-based') methods for statistical analysis; basic understanding of deep learning theory (graph/network theory); and basic general understanding of information technology principles.

Also, the *interested* student will benefit from a basic working knowledge of Python and Keras. Additional working knowledge of one or more of Keras back-end deep learning systems (TensorFlow and/or Microsoft Cognitive Toolkit) will enhance review of included code samples.

NOTE: this course refers to non-graphical analytical methods by this label – 'application-specific' – for want of a better descriptor, though this is by no means standard nomenclature (since machine learning standards remain few and far between at the time of this writing).

Acquirable Primary Skills

Upon completion of this review, the student should acquire improved ability to discriminate, differentiate and conceptualize appropriate implementations of application-specific ('traditional' or 'rule-based') methods versus deep learning methods of statistical analyses and data modeling. Additionally, the student should acquire improved general understanding of graph models as deep learning concepts with specific focus on state-of-the-art awareness of deep learning applications within the fields of character recognition, natural language processing and computer vision.

Again, the *interested* student will also benefit from basic, situational awareness regarding the facility with which these deep learning models may be deployed via Keras and Python using TensorFlow and/or Microsoft Cognitive Toolkit. At the very least, a brief scan of these 'snippets' will help frame and scope the deep learning development process – a store of knowledge that may prove to be of wider benefit during project management planning and budgeting discussions.

About This Executive Review

This text should serve the student as an 'executive review', a distillation of essential information without the clutter of formulae, charts, graphs, references and footnotes. Thus, the student will not have a 'textbook' experience (or expense) while reviewing its contents. Instead, the student will quickly pass through a surprising wealth of actionable, easily-digestible technological information without the distraction of extemporaneous considerations.

Naturally, this objective presents a detriment to the truly research-oriented student, however the associated course is not intended to serve fundamental educational needs. Rather, the information contained herein will rapidly and beneficially update the informed professional regarding use of graph models within deep learning systems via review of historical context, state of the art and, with a bit of extra thought, by improving the student's existing capacity for anticipating lucrative trends within the overlying commercial domain. Best of fortunes!

Foreword

COURSE INTRODUCTION

Course Introduction

About this Introduction

Drawing an analogy from the Cartesian coordinate system, which describes a point in space with three values, this introduction should orient the student within the chaotic space of a new and rapidly advancing technological domain by imparting three levels of insight. The information summarized within this introduction will provide historical background while also describing current state of the art. With these two points of reference in mind, then, the third point of reference, that of the course subject matter, should improve the student's ability to anticipate future trends within this large and vastly influential corpus of technology. Of course, the same information will also prepare the student to derive superior understanding of the subject matter review that follows.

Relevant Nomenclature

To aid the following discussion, several terms first require formal definition. Terms presented in *italicized sans-serif font* may be referenced in the glossary for quick review and definition.

For example, a *parameter*, generally speaking, is a condition that limits or defines performance of a function. A given system may be naturally or artificially constrained to limit the variability of its outcomes by restricting input to the numeric values of one or zero, a state that might be qualitatively described as 'on' or 'off' – these constraints represent the parameters of the system.

Simply stated, within the brief (and painless) overview of statistical methods that follows, the '*dependent variable*' (or '*response variable*') is the 'thing' that the model serves to predict/classify, while the '*independent variable*' (or '*explanatory variable*') is the 'thing' that provides the means of prediction. For example, a model might predict rain according to fluctuations of humidity, which can be readily measured – here, rain is the dependent variable and humidity is the independent variable.

Further, if the response variables trace, more or less, a straight line through the problem space (*i.e.*, rain varies consistently with humidity – which is probably *not* the case), then the model is said to be '*linear*'. Otherwise, the model will be '*non-linear*'.

Finally, '*scalar variables*' provide magnitude (quantity) only. A '*non-scalar variable*', or '*vector*', quantifies both magnitude and direction.

Throughout the text, the term '*system*' predominates. Again, this is a matter of convenience, since many other terms would suffice. In this context, then, the term 'system' references any phenomenon of interest. A system may represent the complex relationships of an entire rain forest ecosystem, but typical systems of commercial interest tend to be much simpler (comparatively speaking, since the Internet is arguably a new kind of forest).

Statistical Inference and Statistical Models

'*Statistical inference*' is the process of producing (or inferring) information about a population using data derived from it. Typically, populations of interest tend to be large enough to prevent direct data extraction from each of its individual data points. This prohibition requires the use of data samples – that is, the process of directly extracting information from only a subset of the population. Given sufficient sample size, the resultant information may be extrapolated to the population at large.

A '*statistical model*' is, in essence, a theory developed to describe some functional aspect of population performance. Application of such models to sampled data supports the identification of properties (or features) that pertain to some relevant aspect of the overall population and/or its dynamics, whatever these may be. In turn, these properties/features should support some meaningful interpretation, description or manipulation (*et cetera*) of the underlying system (*e.g.*, prediction of humidity expected at a specific hour of the day, given daily temperature fluctuations within the region).

Contrasts with Descriptive Statistics

In contrast to statistical inference, '*descriptive statistics*' use similar methods as applied to similar datasets, albeit with an alternative desired outcome. As its label implies, descriptive statistics provide a means of quantitatively describing or summarizing the population's various aspects (as opposed to inference, which additionally uses these results to provide some practical insight into the problem domain).

Lacking a predictive requirement, descriptive systems do not rely on probability theory – rather, these tend to be nonparametric problems that provide 'details' useful for understanding the underlying phenomenon.

As a practical example, a scientific experiment might use inferential statistics to draw its functional conclusions – *e.g.*, efficacy of dosages appropriate for a patient by gender and age. The same experiment might also use descriptive statistics to provide the context of study applicability – *e.g.*, statements that describe the target population of the study as being restricted to male patients more than sixty-five years old ("gender" and "age" being the descriptive statists).

Modeling Assumptions

To eliminate uninteresting variabilities and biases within an analysis, modelers typically attempt to simplify the underlying system by discarding irrelevant data (*e.g.*, an analysis attempting to predict a stock price might not benefit from input that describes lottery ticket sales). Another necessary aspect of these considerations must regard the logical assumptions used to qualify application of a given statistical model (or method), since these assumptions support simplification at the cost of limiting applicability.

For example, where a model or method assumes a normal distribution of data (*i.e.*, the 'bell curve') and is also dependent upon a random sample extracted from same (both common assumptions of many regression-based protocols), then any attempt to apply the model to a distribution that is *not* normal and/or *not* drawn from a random sample will produce invalid or biased results. In part due to these considerations, for example, fully parametric models typically do not produce satisfactory results when applied to samples drawn from human or economic populations (unless these samples are extremely large), since the assumptions of normalcy and randomness are often inappropriate.

Three Levels of Modeling Assumptions

'*Fully parametric models*' include a 'family' of distributions (mathematical functions providing probabilities of occurrence of various potential outcomes) described by a finite number of known parameters. For example, linear regression methods generally apply as fully-parametric models, since they provide accurate results when these assumptions are true.

'*Non-parametric models*' include distributions with an infinite number of unknown parameters. Human and economic behavior systems provide adequate examples here, since any attempt to predict future performance based on past performance using parametric models will provide unsatisfactory results, largely due to invalidation of assumptions about the underlying dataset.

'*Semi-parametric models*' are simply hybrid solutions. These models may contain distributions demonstrating both finite/known and infinite/unknown parameters.

Simply described, within parametric solutions, the student develops the model structure first (*e.g.*, a regression line), and then applies it *to* analytical datasets amenable to the required assumptions. Contrarily, within non-parametric solutions, the model, itself, derives *from* the underlying dataset.

Indeed, this is the primary course focus, which reviews contemporary solutions to complex problems that typically cannot be constrained to meet assumptions of simplicity. Stated another way, both types of models have *parameters*, the difference between them primarily being the number and nature of these parameters – parametric models have a limited number of fixed constraints, while non-parametric models have an infinite number of variable constraints.

In practice, within the limited realm of valid applications, parametric models produce more accurate results, but at the cost of (often) crippling restrictions. These solutions traditionally present 'application-specific', 'hard-coded', 'rule-based', manually derived formulae for analytical purposes. In practice, too, these models become less useful as the problem space approaches 'real life' conditions (thus, invalidating model assumptions). As applied to these more complex systems, then, 'application-specific', 'hard-coded', 'rule-based' solutions become less and less viable, largely due to considerations of (unrealistically protracted) compute time.

Until the advent of powerful computers combined with recent innovations in graph modelling theory (and, eventually, 'deep learning' software), parametric models provided simplified, rapidly computable methods of statistical inference, albeit with limited utility. Today, in the age of 'big data', parametric models suffer from reduced scalability and applicability and, so, cannot be effectively applied to the resolution of many pressing physical, medical and social phenomena.

Application-Specific Methods

'*Application-specific*' ('*hard-coded*', '*rule-based*') methods of statistical analysis are common and useful solutions to many practical problems, especially among datasets and problem spaces that support parametric models. Unfortunately, these systems become less robust and useful as the size and complexity of the dataset increases, especially where these data derive from chaotic sources (*e.g.*, real-time web traffic, economic trends, *etc...*).

Because these models rely on simplifying assumptions, resultant solutions tend to be highly qualified as to their efficacy – valid results are applicable only where underlying assumptions prove true. Where these solutions suffice, however, computation of results is an efficient, useful prospect. Typically, the necessary develop paradigm is also comparatively simple (One Problem -> One Solution/Model -> One Algorithm).

Again, while combinatorial approaches may compensate for parametric limitations, the resultant aggregated solution is too often irrationally complex and prohibitively costly, especially when applied to contemporary 'big data' problems. As a means of illustrating the utility, and limitations, of application-specific models, this course briefly reviews common examples of these methods.

Examples

Most first year college students will recognize the most popular variants of the '*linear model*' (*i.e.*, '*linear regression*'). A partial list of examples includes *ANOVA* (ANalysis Of VAriance), *ANCOVA* (analysis of co-variance), *MANOVA* (multivariate analysis of variance), and *MANCOVA* (multivariate analysis of co-variance). Most often, these students will also use the '*T-test*' and '*F-test*' of significance within their first-year science coursework – this information derives from an underlying regression analysis.

As previously stated, linear models typically require multiple assumptions (*e.g.*, assumption of normalcy and randomness). Consequently, novices often misapply them to datasets that violate required assumptions, thus producing misleading outcomes. Further, a simple linear regression models/predicts only a single "*scalar response*' (or dependent variable) to a single explanatory (independent) variable. Practical systems are seldom so simple – hence, the increasing complexity of the examples listed below, which extend simple regression to analyze for multivariate and covariate responses.

The '*multiple linear regression*' model further extends the application of these methods. As the label implies, these models perform the same simple linear regression, albeit with multiple explanatory (independent) variables. Similarly, '*multivariate linear regression*' models predict more than one correlated response (dependent) variable.

Another form of regression model analyzes non-linear systems, or systems wherein the dependent variable does not change at a unified rate – thus, forming a curved line, as opposed to a straight line. Unsurprisingly, these are called '*non-linear regression*' models, which is also a form of '*curve-fitting*'. Examples of such systems include exponential-, logarithmic-, trigonometric-, power- and Gaussian functions, as well as Lorenz curves.

While a detailed discussion of these non-linear methods extends outside the scope of this course, certain aspects of their application bear minor mention here. In a sense, non-linear systems begin to approach a 'landscape' of distributions – in other words, within multivariate systems, these multiple non-linear models, stacked together in two-dimensional space, began to describe relationships in three-dimensional space resembling mountainous terrain that includes multiple 'valleys' and 'peaks'. These '*minima*' and '*maxima*', respectively, describe the lows and highs of the response variable(s) as the explanatory variable(s) change values.

As any mountain hiker knows, the low point of a single valley may (or may not) also be the low point of the entire range (or continent, for that matter). This relates to the concepts of *local minima* versus *global minima* (or, conversely, *local maxima* versus *global maxima*) – while local minima/maxima may provide practically useful outcomes, these are not necessarily the *optimal outcome*. Instead, the optimum value typically resides at the global, or overall, minima/maxima.

Consequently, these models suffer from bias attributable to selected 'starting points', or variable initialization, so they often require '*numerical optimization*'. This is simply the process of choosing, perhaps randomly, multiple estimations of initial parametric values to more completely explore the problem space and correctly identify these global values of interest.

Returning to the previous analogy, this process resembles the placement of multiple hikers within the landscape, each tasked with reporting *local* values with the goal of performing a post-analysis procedure to identify *global* values (or, at the least, best approximations). Unfortunately, only an '*exhaustive analysis*' guarantees location of the *true* global minima/maxima – this is the equivalent of placing one hiker near each valley/peak, a prospect that quickly expands beyond the point of practicality when the landscape is vast.

As implied by the label, a '*logistic regression*' typically applies to *binary variables*, which have the value one or zero (*i.e.*, off/on, yes/no, true/false, *etc*...). These models are, essentially, a standard linear regression employed where the log-odds probability of a phenomenon is a linear combination of independent variables.

As with standard regression, these models generalize to multiple (more than two) levels of the dependent variable. '*Multinomial logistic regression*' supports analysis of unordered categorical outputs with more than two values, while '*ordinal logistic regression*' applies to similar systems of ordered categorical outputs.

'*Decision tree*' analysis provides another means of analyzing complex systems that are inherently chaotic (at least, to some significant degree). Originally developed to visualize algorithmic flow (*e.g.*, visualization of multiple, branching IF/THEN/ELSE logic), these methods use a tree-like graph/model to map all the decisions thus produced, a concept revisited often throughout the review that follows.

Currently, these tools find a home in analyses used to determine complex, interacting decision-pathways most likely to produce a goal of interest. Within the context of this discussion, these rudimentary graph-based models provide a convenient segue into its primary subject matter, machine learning.

Predictive Modeling and Machine Learning

'*Predictive modeling*', or '*predictive analytics*', overlaps with the machine learning domain to the point of synonymity. As with the application-specific methods previously described, machine learning techniques attempt to provide practical insight into complex systems via application of a simplified model of the system of interest. Largely due to advances in computational hardware and software, machine learning models now present viable solutions to many otherwise intractable analytical problems. Specifically, these 'learning systems' rather conveniently provide viable solutions to the complex 'big data' problems made available for investigation by increasingly capable data-storage (and processing) hardware.

In lieu of application-specific, rule-based algorithms, these AI systems accomplish this goal by autonomously 'learning' the parametric relationships of the model using input from the dataset of interest. Typically, analysts use these '*artificially intelligent*' (*AI*) solutions to predict or classify complex, multivariate phenomena that are not amenable to the solutions previously described.

Predictors and Classifiers

While the terms 'predictor' and 'classifier' bear no formal definition here, a brief review of specific implementations of these methods will provide a 'backstory' useful to understanding the need (and the 'hype') attributable to resurgent interest in all things AI. Of particular note, evolution of the following, largely application-specific, examples will illustrate the increasing... angst... with which modern statisticians confronted 'big data' problems at the turn of the current century (ponder Google's fantastic rise using amazingly responsive web-search analytics).

Accordingly, a brief, progressive description of several examples of interest will illustrate the growing analytical 'frustration' (for want of a better word) that the use of graph-based models (AI) now promises to resolve.

Examples

Closely related to standard regression models previously described, '*Generalized Linear Models*' *(GLM)* generalize linear regression by relating response variables via linking functions and by making the magnitude of each measurement's variance a function of its predicted value. Already reviewed, logistic regression falls into this family of tools, since its underlying use of function-linkage provides similar utility. Likewise, '*generalized additive models*' are generalized linear models wherein linear predictors depend linearly on predictor variables that are, themselves, representative of unknown smooth functions. These functions, to some extent, incorporate 'tolerance of randomness' into model design.

Another attempt to apply application-specific analytics to complex problems provides a means of simple regression that deals adequately with outliers while also providing moderate insensitivity to violation of model assumptions (randomness). This process is called '*robust regression*', which comprises a family of solutions to the same problem domains that benefit from machine learning approaches now dominating the industry. Closely related, '*semiparametric regression*' attempts to resolve similar deficiencies with similar results by employing combinations of parametric and non-parametric models. Most students will not use these techniques, largely due to the overwhelmingly popular utility of AI (specifically, deep learning) methods.

'*Ordinary Least Squares*' represents another attempt to resolve these difficulties using, unsurprisingly, '*least squares methods*'. This technique is a form of '*data fitting*', wherein the best fit (solution) of a model minimizes the sum of squared residuals (the difference between observed values and fitted values, as provided by the model). As with all similar techniques, these methods suffer to some degree from restrictive assumptions and over-simplification of data due to analytical compromise.

'*Random forest*' and '*boosted tree*' methods provide a higher-order implementation of the decision trees previously described. A form of '*ensemble learning*', random forests provide a means for classification, regression and other analytical operations executed against large, multivariate data. These protocols construct a '*forest*' of decision trees via '*training*' (exposure to a subset of the data) to output the mode class (in classification problems) or mean prediction (in regression problems) of its individual trees. Like many 'trainable' algorithms, an incautious analyst can inadvertently *overfit* the training data.

'*Overfit*' is a common problem in machine learning domains. This is the process whereby a graph-based model excessively or inappropriately trains on a subset of the data to the detriment of its performance when applied to the target dataset. Think of a glove designed to fit one mannequin's hand so well that it no longer generalizes to fit the wider population of intended consumers.

'*Naive Bayes*' is a 'family' of '*probabilistic classifiers*' derived from '*Bayes' theorem*' (essentially, a set of equations for managing stochastic, or random, phenomena) – though purists will argue these are not truly Bayesian applications within the context of machine learning. These methods assume robust independence among feature values.

Graph Models for Deep Learning

The '*k-nearest neighbor*' *(k-NN)* algorithm is a non-parametric method used for classification and regression. When used for classification, the output is a class label derived by majority 'vote' polled among the neighbors of a given input – *i.e.*, the output class for a given object will be the class assigned to the majority of its nearest neighbors (as determined by its 'position' within the multivariate analytical domain). When applied to a regression problem, k-NN output is the property value of the given object.

k-NN relies on '*lazy learning*'. This process only approximates the classification model locally while deferring computation until classification. This is often considered to be the simplest machine learning algorithm.

'*Support Vector Machines*' (*SVM*) ('*non-probabilistic binary linear classifiers*') are supervised learning models that also perform classification and/or prediction (regression). These methods require the implementation of training algorithms to support the learning process, which proceeds by marking each training example as belonging to one of two possible classes. Upon completion of analysis, the two classes should map to points in space divided by a clear gap structured to be as large as allowable. Once trained, new data enters the classifier to fall into one of these identified classes.

An '*Artificial Neural Network*' *(ANN)* is biologically-inspired computational system designed to '*learn*' tasks (such as prediction and/or classification) by first examining a labeled subset of the unlabeled data. An ANN is, essentially, a collection of '*nodes*' (*artificial neurons*) that connect to one another in a variety of pre-ordered patterns along '*edges*' (the '*artificial synapse*'). During operations, each node in the system receives input, performs some operation on this input, and then passes this information as output – the sender/receiver of these arrangements (the architecture of the network) is very much a matter of continuing development, as well as a central subject matter of this course.

Graph Theory and Neural Networks

Not to be confused with the visual rendition of data that often takes the form of squiggly lines or blocks of colors, within the context of this discussion, 'graphs' are mathematical constructs used to model relationships between objects. As with the neural networks of biology (and your brain), graphs consist of two simple concepts – nodes and edges. Within a graph, nodes (also called '*neurons*', '*vertices*' or '*points*') connect to one another via edges (also called '*arcs*', '*lines*' or '*synapses*') in a dizzying array of potential patterns.

Graphs may use a '*directed*' or an '*un-directed*' architecture to describe its arrangements of nodes and edges. Within a '*directed graph*', information flows one-way, especially when the graph contains multiple layers (*i.e.*, nodes can only send *or* receive, exclusively, from a given edge). Conversely, an '*undirected graph*' supports bi-directional transfer of information (*i.e.*, nodes may both 'send' *and* 'receive' information from all of its edges). Within visual depictions of the graph, arrows indicate its '*directedness*' (as opposed to un-directed graphs, which depict edges as straight lines).

Without immediately confounding graph models with neural networks, artificial intelligence and deep learning (which follows in detail), a quick review of basic graph models as viable practical solutions will provide guiding context to the larger discussion that follows. Accordingly, briefly review examples of these tools in contemporary settings.

Applications

As applied to '*linguistic analysis*' – the analysis of language – graph models broach the capacity of many application-specific methods previously discussed to provide efficient and efficacious means of analyzing for discrete structures found within language systems. The tree- and forest-based methods described above proved most useful in this regard, since human language readily lends itself to this kind of dissection. Indeed, many systems perform according to similar constructs, so applications in linguistics (as with other fields of study) transcend technological boundaries to find utilities within many different domains.

Graph models also find abundant applications within the natural sciences. A definitive list of these tools quickly expands beyond the scope of this effort, but a summary review of these implementations reveals graph models working happily within physics laboratories when quantitatively modelling complex, three-dimensional arrangements of atoms within material topologies. These same techniques proved useful within biological labs tasked with parsing and identifying complex molecular structures, ecological networks and epidemiological phenomena, among many more.

Social sciences successfully use graph models to explore complex social phenomena such as interpersonal relationships, social networking, information spreading and more. Indeed, as previously discussed, these 'human-oriented' problems exemplify optimal target domains for application of these machine-learning concepts.

Mathematics and computer science applications abound, as well. Indeed, graph model theory advanced considerably when applied to information technology problems that include topological studies, network router optimization, internet link mapping and countless additional examples too numerous to adequately embrace here.

Finally, graph models provide the basic component of neural networks. As these networks evolve into artificially intelligent, so-called deep-learning constructs, the power of graph models revealed itself in a record number of successes accumulated through the last few years. Many of these achievements resolved previously intractable problems that often proved the limits of the application-specific models previously described. In essence, these graph-model-based systems work because they 'learn' their functionality without requiring the intensive human effort associated with 'hard-coding' impossibly complex rule-based algorithms.

Graph Models and Deep Learning

As implied by the label, 'deep learning' systems surmount and overcome the limitations of application-specific, hard-coded or rule-based algorithms by 'learning' trends or features within a dataset. These increasingly complex implementations achieve results via application of rationally designed '*cascades*' of graphs – that is, multiple graph models working together within a layered hierarchy, wherein each successive layer receives its input from the preceding layer's output. In these implementations, the top layer is the initial system input layer and the bottom layer provides the system's final output.

Deep learning networks have proven useful within the domains of prediction and classification, much like the application-specific methods discussed earlier. Their increasingly capable utility derives directly from the processing of training (learning) used to 'teach' underlying data '*features*' (or useful, diagnostic characteristics). The 'traditional' converse of this process, loosely labeled '*automated feature extraction*', is called '*feature engineering*', which is, essentially, a fancy term for describing the process of *manually* developing an applicable model, as with the application-specific algorithms of the introduction. Feature engineering is the single costliest aspect of these traditional methods, both in terms of money and manpower, and this is the primary advantage to the use of artificially intelligent, machine-learning solutions.

Relevant Nomenclature

Within the context of this course, a '*feature*' formally represents an individual, measurable property or characteristic of an observed phenomenon or object. These are, precisely, the aspects of any model that render its utility for prediction or classification, when applied to a dataset of interest.

Again, as the label implies, the *'learning'* process required prior to deployment of a viable deep learning model requires *training* of some kind. As with the variety attributable to graph model architectures, many forms and methods of training exist, each tailored for a specific model architecture, problem or both. Indeed, the least efficient and most time-consuming aspect of deep learning implementations is this process – the process of training. Depending upon the details of a given implementation (especially as regards the quantity, quality and availability of adequate training data), these learning methods are, themselves, a primary source of many recent innovations.

Furthermore, training of this nature may be *'supervised'* or *'unsupervised'*. As the student might expect, supervised training requires *'labeled training data'*, which requires some form of preliminary automated or manual intervention, as needed to mark the training dataset. Conversely, unsupervised methods do not require labeled training data. The choice of methodologies depends, in large part, on the type of data under study, since some data are easier to label/annotate than others (*e.g.*, textual data in a delimited flat file containing one million records is much easier to label than are the outlines of every human figure found within one million photographs).

Applications (Graph Models)

Although this course will not detail the many existing, practical deployments of this technology to resolve 'real-world' problems, the professional student will benefit from a brief review of key fields within this domain, since many will originate from, or be destined to serve, these technological demands as gainfully employed professionals (or well-meaning amateurs).

Since graph-models excel as solutions to complex, often chaotic multivariate problem spaces, the following summary list of breakthrough applications will further enlighten the interested student regarding the lucrative interests currently driving advances in this huge and expanding marketplace. Briefly tagged, these fields of research and innovation include bioinformatic drug design, near-human expert gaming agents, natural language processing (NLP) for intelligence collection, social network and message filtering, machine translation of auditory and visual language input, speech recognition and transcription, audio recognition and reproduction, computer vision and many, many others.

Clearly, graph models present a hidden powerhouse of potential, as realized in these machine learning algorithms. Obvious aspects of utility no doubt drove the individual student to seek this course and others like it, since variations of these tools will explode onto the technological problem space in the months (perhaps years) to come.

Applications (Deep Learning)

Without digressing into the abundant fine detail easily attributable to even the most cursory review of deep learning designs, this course will provide executive-level granularity within the following descriptions of deep learning technology. Graph models play a central role within these solutions, but contemporary renditions of this component vastly expand and enhance the essential functionality of these concepts through an ever-increasing outlay of new deep-learning constructs and implementation paradigms.

Popular base implementations of these solutions include deep neural networks, deep belief networks, recurrent networks and convolutional networks, among others. These variations intermingle and combine within truly deep networks (hence the term 'deep learning') in an abundance of configurations, each with its own best practices of development and applications.

Lesson Conclusion

An executive-level review of the subject matter described within this course will generally inform the student regarding historical context and state-of-the-art in deep learning concepts. Specifically, this course should inform the student regarding the role of graph models within the realm of deep learning implementations.

This introduction briefly reviewed relevant application-specific algorithms and 'traditional' rule-based, often hard-coded, solutions for classifying or predicting values from complex, multivariate datasets. Specifically, discussion focused on the appropriate and, more importantly, inappropriate application of these models to large, diverse, sparsely populated datasets of the kind commonly encountered within the contemporary 'internet of things'.

Where these (often highly) constrained parametric models begin to fail, graph-based models become viable solutions. In turn, as with all things technological, these simple models expanded horizontally and vertically from simple Support Vector Machines through Artificial Neural Networks into the deep learning constructs popular today.

Building on these concepts, then, this course will provide *general* insight into the deep learning domain while *specifically* focusing on the role of graph models within these systems.

Describing Model Structure with Graphs

Lesson Overview

Recall from the preceding introduction that graphs are mathematical constructs that model relationships between objects of interest. These graphs consist of *'nodes'* (*'neurons'*, *'units'*, or *'points'*) connected to one another via *'edges'* (*'dendrites'*, *'synapses'*, *'arcs'* or *'lines'*). Information passes from one node to another via edges, and each node performs some operation on the information (or not, depending upon the nature of the graph) before passing it to the next. In this sense, graphs may be directed or undirected, wherein information moves along one-way paths (as indicated visually by arrows) or two-way paths (visualized as lines).

Relevant Nomenclature

As with many technological topics, understanding of lesson objectives will improve with a brief review of relevant nomenclature. Within the context of this course, the term *'adjacent'* or *'adjacency'* indicates two nodes that share a common edge (or, alternatively, two edges that share a common node, depending upon perspective). The term *'incident'* or *'incidental'* refers to an edge and a node on that edge (or a node and an edge on that node, again, depending upon perspective).

Incidentally, graphs often present as visualizations – that is graphical depictions of the nodes-and-edges arrangement of the model's structure. While these graphical renditions appear graph-like, they are not, as presented within this discourse, the graph, itself. Instead, these models are, in practice, abstract mathematical or computational constructs that exist only *in silico* (within the computer) or as formulae written on paper. This, then, is the *actual* graph – that is to say, the object of most discussion that follows.

Logical Representation of Graph Structure

As previously stated, the functional graph-model will be an *in silico* or paper-based construct that performs the actual operations of the implementation. These logical representations must preserve vital information about the graph, including descriptions of its layers, their nodes and edges, connectivity and other details discussed in lessons to follow. While many methods for displaying and operating a graph model exist, this course focuses on *in silico* modes of representation.

Consequently, this section of the course describes common methods for computationally implementing graph models of various kinds. While detailed design and delivery of these products exceeds the scope of this course, the student will benefit from a summary review of common logical constructs employed to provide functional models within active solutions.

List Structures

'*List structures*' describe the graph as a collection (or array) of '*unordered lists*' (a common algorithm development paradigm presenting a wide array of organizational possibilities not detailed here). List-based implementations optimize for performance of sparse graphs in systems possessing limited memory. Most often, these lists are, in fact, *lists-of-lists*.

Lists may describe adjacencies or incidents. '*Adjacency lists*' describe each node's edge set. Conversely, '*incident lists*' describe each edge's node set. These constructs are lists-of-lists, in the sense that the top-level list contains sub-lists, which are the descriptive-level component of the system – each sub-list describes the *node/edge set* of a single edge or node, respectively. Design-time selections of these alternatives is need-specific. Combinations are also common.

Finally, a minor word about optimization. Given the node- or edge-based focus of adjacency *versus* incident lists, each one offers a specific form of access to model components. Since adjacency lists are node-focused, unsurprisingly, these methods provide handy access to a model's '*node collection*'. Conversely, again, incident lists provide similar access to a model's '*edge collection*'.

Matrix Structures

While list structures optimize for sparse graphs and/or systems with limited memory, these simple constructs are not necessarily the overall optimal choice for deep networks. Instead, given the multi-layered (multi-dimensional) nature of these complex neural networks, '*matrix structures*' may support superior performance. Again, these concepts often combine such that a given implementation may, in fact, contain a list-of-matrices, rather than simply containing one large, monolithic matrix that describes the entire model, layer by layer (where each sub-list describes a single layer's structure).

Matrix-based implementations of graph models optimize performance of dense graphs at the cost of increased memory requirements. As with list-based constructs, matrix descriptions may provide either adjacencies, incidents or combinations of both. At sub-list level, matrices typically project edges by rows and nodes by columns ('*adjacency matrices*'), or nodes by rows and edges by columns ('*incident matrices*'). Each matrix entry (or cell) contains either a zero or one to indicate node-edge connectivity, where one indicates a connection and zero indicates lack thereof.

Implementation

While list and matrix structures provide essential implementation of graph notation, they are far from the only means of deploying such systems. The student will benefit from a brief review of key concepts within this realm of study.

'*Hash table*' implementations leverage the versality of hash functions to store and access graph model components. A hash function is simply an algorithm structured to ensure one-for-one storage and retrieval within a non-indexed, non-hierarchical container. Typically, these solutions encode each node as a hashable object of any kind, while storing the graph collection as a simple array. Hash table products may not explicitly encode edge information.

Indeed, array constructs are common. The basic modes of implementation previously discussed, lists and matrices, are, essentially, arrays of various kinds. In fact, more complex graphs typically deploy as arrays-of-arrays (theoretically described as lists-of-lists or lists-of-matrices) in any combination that meets design criteria.

Simple arrays typically reside in memory as relatively monopartite constructs (*i.e.*, a fixed memory container of given size filled in some rational order). More complex arrays support more complex interactions and operations.

'*Indexed arrays*' are simply arrays implemented with some kind of index. This index is just a number (or some other kind of logical handle) that identifies a unique entry or slot within the array container. Within this approach, the top-level array contains the graph's nodes (or edges), while each array entry represents another array (or list or matrix) that describes the node's neighboring edges (or the edge's neighboring nodes). This general situation describes directed graph models. Undirected graph models will present top-level arrays that contain two sub-arrays (or lists) for each item to describe bidirectional connectivity (one for each direction).

'*Object-oriented list*' constructs are also common targets for encoding complex deep-learning models. These implementations include custom-designed logical objects that incorporate a variety of operational support for the graphical model in storage. Or not, depending on design criteria. Again, developers typically take an adjacency, incidence or hybrid approach to architectural design. While this option offers the most flexible methodology, it also requires programming expertise. Fortunately, a variety of open source (and proprietary) applications do this well. A partial list of popular variants will include Caffe, TensorFlow, Microsoft Cognitive Toolkit and many others.

Examples

Naturally, most commercial products provide more than simple containers for storage of graph components. Graphical models of various kinds are available from many vendors in many formats targeting a variety of problems. These higher-level solutions typically encode the full functionality of a particular graph, thus offering an all-in-one package for implementing the graphical model of interest. The student will benefit from a brief review of popular iterations of such applications.

'*Factor graphs*' are undirected bipartite graphs that connect objects called '*variables*' and '*factors*'. Each *factor* represents a function over its associated *variables*. These somewhat specialized tools support implementation and analysis of belief networks (propagation of beliefs within a population), among others.

'*Clique trees*' ('*junction trees*' or '*join trees*') provide another specialized tool, in this case used to perform tree decomposition (graph-to-tree component mapping) to determine the '*treewidth*' of the graph. This information, in turn, supports analysis and optimization of graph model performance. These tools also support probabilistic inference, query optimization, and matrix decomposition.

'*Chain graphs*' provide a method for unifying and/or generalizing Bayesian and Markov networks (detailed below). These models have both directed and undirected edges without directed cycles. A '*directed cycle*' is described as a pathway through the graph that starts at a given node and then follows a single, unique pathway through that node's directed edges to ultimately return to the originating node – a sort of graphic loop. Special cases of chain graphs include '*directed acyclic graphs*' and undirected graphs.

'*Ancestral graphs*' provide tools for studying graphical models, themselves – this is, in fact, an endeavor of growing interest, given the increasing complexity of deep learning networks. These are '*mixed models*', since they contain directed, undirected and '*bidirected edges*'.

A '*conditional random field*' is a type of discriminative, undirected probabilistic graphical model that encodes relationships between observations to construct interpretations of these relationships. Several common applications described in subsequent sections of this course benefit from these solutions, including Natural Language Processing (NLP), biological sequence analyses, molecular network analysis, and computer vision.

'*Bayesian networks*' ('*directed graphical models*', or '*belief networks*') are '*Directed Acyclic Graphs*' *(DAG)* that support probabilistic graphical models representing variables and their conditional dependencies. These systems output probabilities related to the likelihood that a phenomenon of interest will occur within a given system. Special cases of Baysian networks include '*Hidden Markov Models*', neural networks, '*Variable-Order Markov Models*' and others.

Lesson Conclusion

Analysis, development and implementation of graph models naturally requires some method of visualizing its architecture, but this visualization should not be confused for the functional graph, itself. These are logical (or formulaic) constructs that perform the actual operations attributable to model functionality.

Developers encode these logical constructs in a variety of ways, which generally evolved from list and/or matrix structures. The former optimizes for sparse graphs performing in limited memory environments, while the latter optimizes for dense graphs at the cost of additional memory requirements.

A variety of popular open source software now implements these systems, largely negating the need for customized development of core functionality. Many contemporary applications simply build on these backend services to perform a wide and growing range of analyses with the models thus produced.

As might be expected, this rapidly expanding technology has a long, diverse and ever-growing list of 'real-world' applications. Within biological research, graphical models impact the analysis of regulatory genetic networks, protein structure prediction, protein-protein interactions, protein-drug interactions, free energy calculations and countless others. Information and intelligence research benefits from applications to causal inference, information extraction, natural language processing, multimedia data processing, speech recognition, computer vision and more. This course reviews key examples of these concepts in the sections that follow.

Deep Learning and Graphical Models

Why Deep Learning? Why now?

Likely, the student has not undertaken his course as a matter of random chance. Given the large, growing and lucrative demand for machine-learning professionals, most students will complete this course to improve their functional understanding of a hot topic while expanding and enhancing their professional acumen and credentials. Accordingly, the interested student will benefit from a brief review of recent history, as summarized within the evolutionary timeline of this promising technology.

While the concept of logical (or artificial) neural networks have existed since the dawn of commercial computing, limitations of hardware largely restrained these ideations to the realm of theory. In fact, the term '*deep learning*' (or '*Deep Neural Network*' [*DNN*]) first appeared within scientific literature as early as 1986, long before hardware could support its practical implementations.

During this period, however, something magical happened to the business of information technology. Video games moved from glitzy (and oftentimes expensive) commercial arcades into living rooms around the globe. As this entertainment powerhouse advanced in popularity and profits, a growing consumer demand for faster, more powerful and more capable graphical displays induced vast expenditures of research and development investments.

Computer displays subsequently advanced from text-based monochromatic presentations (usually bright green or orange!) through the '*Color Graphics Adapters*' (*CGA*), '*Video Graphics Adapters*' (*VGA*), '*Super-Video Graphics Adapter*' (*SVGA*), '*Digital Visual Interface*' (*DVI*), '*Video In Video Out*' (*VIVO*) for *S-Video*, and '*High Definition Multimedia Interface*' (*HDMI*) standards. Simultaneously, specialized, 'heavy-lift' computational demands of the increasingly intensive three-dimensional graphics and multimedia gaming experience shifted from the CPU (Central Processing Unit) to specialized cards designed to present 'fat' numerically-based data pipelines that streamlined the production of graphical content straight to the user through his or her peripheral devices. These 'Graphical Processing Units' (GPU) revolutionized the gaming business, virtually overnight.

In 2009, Nvidia, a leading vendor of these graphics chips, adapted their popular GPUs to support neural network training routines. That year, the Google Brain team used these devices to create capable DNNs, improving training speeds by two orders of magnitude (100x). Essentially, that year, hardware designed to efficiently support the matrix- and vector-based math used by 3D graphics software to manipulate 'virtual objects' in 'virtual space' met graphical model theory, which happened to be driven by various implementations of... drumroll, please... matrix- and vector-based mathematical constructs.

Following this success in 2011, the Google Brain team developed a proprietary deep-learning system they labeled 'DistBelief'. Benefits of these, now practically useful, models quickly proved profitable and promising, driving wider adoption and development of machine-learning applications within the commercial search-engine enterprise – quickly spreading from there into most technological domains within the span of a few years (months, in some cases).

In part, this explosion of applications is due to a 2012 presidential 'Big Data Research and Development Initiative', which guided commercial, academic and public-sector researchers to contemplate big-data problems and their solutions. Finally, in 2015, Google released 'TensorFlow' under the Apache 2.0 open source license, thus providing free, open-source deep-learning software to a ravenous demand. Others quickly followed.

Lesson Overview

Defined in simplest terms, '*deep learning*' (*DL*) models are basically 'souped-up' graph models in the guise of a deeply layered neural network. Hence, the 'deep' part of its label.

These are machine learning constructs with a specific architecture (though the *archetype* [essential nature] of this architecture remains somewhat nebulously defined – in terms of standard, at least). As a definitive aspect, these deep-learning systems rely upon a '*cascade*' of nonlinear processing units (neural networks). As with most neural nets, the output of each layer provides input to subordinate layers, thus the 'cascade' of operational flow.

As with most graph models, these solutions deploy to perform feature extraction and transformation of complex, multivariate datasets. Also, the training (or learning) process may be performed in a supervised (*e.g.*, within classification roles) or unsupervised manner (*e.g.*, within pattern analysis roles). Finally, these neural networks learn multiple levels of data representation, which are often – but not strictly – related to individual network layers within the system.

The overall structure and the layer-wise architecture of deep-learning models present many computational challenges, primarily as regards the training process. Prior to the advent of multi-core CPUs, hardware/software accelerators and GPUs, this crucial aspect of deep-learning implementation presented an intractable hurdle that left many interesting problems unresolved.

Indeed, optimization of training performance continues to challenge investigators as larger and more complex datasets become more accessible to graph model analyses. The student will review the most relevant examples of these

Graph Models for Deep Learning

systems in successive sections of this course, but a few examples bear minor introductory mention here, such as 'Deep Belief Networks' and 'Deep Botzmann Machines'. Both are popular theoretical designs of deep learning that have proven themselves effective against many large real-world problems.

Relevant Nomenclature

Throughout typical discussions regarding design and construction of neural networks, the term *'layer'* appears often. While the overt meaning of this word applies in the strictest sense to basic deep neural network components (*i.e.*, individual layers of the graph model), these layers may be of two specific kinds. *'Visible layers'* accept initial input into the model and, alternately, present its final output – these layers are 'visible', or accessible, to the user or developer. *'Hidden layers'* represent the *latent variables* of the model, occupying its central, functional core, which is not accessible to the user or developer. Simply put, the visible layers 'get all the glory' while the hidden layers 'do all the work'.

Essential Deep Learning Concepts

Deep learning (DL) systems differ from the neural networks (and other 'basic' machine learning methods previously discussed) according to the depth and breadth of the underlying architecture. Deep learning systems benefit from their large *'Credit Assignment Path' (CAP)* depth. CAP is basically a stepwise count that denotes the chain of data transformations within a given system, progressing from input to output. While no hard standard defines the formal separation of 'shallow' neural networks to 'deep' neural networks, the *de facto* accepted standard describes deep systems as having a CAP depth greater than two.

While a brief review will illuminate many benefits and advantages of these solutions as applied to complex, multivariate datasets, their most essential benefit relates to a reduction of human intervention (thus, costly man-hours of labor). Essentially, these systems circumvent the need for teams of engineers focused exclusively on traditional methods of development using application-specific models derived from manually engineered features.

Understanding Deep Learning

Occasionally, a less-informed technology professional may opine that nobody 'truly understands' the process by which deep learning systems learn. This is a misstatement. While some debate remains regarding the specifics of this process (as with the functionality of real, biological neural networks), two schools of thought currently present viable theories regarding its essential nature.

Universal Approximation

The '*universal approximation theorem*' identifies the practical functionality of DL systems with their capacity for approximating continuous functions as derived from the architecture of a basic feedforward neural network (NN) having a single hidden layer of finite size. A crude analogy will illustrate this concept.

Imagine a fishing net stretched onto a beach such that its web forms a regular mesh of rectangles – if each of the net's knots represents a node and the lines between knots represent edges, then the net adequately represents an uninitialized, fully-connected NN having a single layer of finite size. Next, imagine the fisherman laying a large a large flat plane beneath the net – perhaps a bit of plywood – with data points marked on it in red paint (these points appear random on the board, though the fisherman suspects the existence of useful patterns within the mixed and anonymous dots). To identify this pattern, the fisherman paints one half of net-knots blue and the other half green and then, finally, he begins to arrange the knots of the net, one at a time, over the points on the board that he believes represent membership in one of the two classes he wants to reveal (you may recognize a tree, forest or nearest-neighbor approach in this method). The fisherman then proceeds to twist, tug, pull and arrange the net's fabric such that its two colors, blue and green, overlay the dots on the board to correctly classify original red points into one of the two groups, blue or green, thus 'training' the net to perform a desired classification. Now the fisherman can 'freeze' the form of the net and then use it like a rigid template to rapidly classify the red dots of another, similar board.

In this case, the 'frozen net' is the final functional graphic model and this process crudely resembles the concept of 'universal approximation'. By twisting the net about until its once-straight knots-and-lines conformed to the natural, contorted (chaotic) boundaries of the dataset (red dots), the fisherman approximated unknown parameters of the system to produce his working template (model) based on his arrangement of its layout (training). Incidentally, this analogy requires the lines between knots to vary in length (perhaps the net is elastic), and a measure of this variable distance represents the final, trained value of the network's nodes – it is, precisely, these re-weighted values that constrain the net to encode model parameters, thus producing a viable classification gap between samples.

Finally and briefly, to imagine how this process extrapolates into deeper designs, imagine these nets stretched vertically, as well, to describe highs-and-lows among the data points. Further, expand the example down the beach, using multiple nets and multiple landscapes. Then, somehow, bring them altogether in a single space to provide a robust means for simultaneously mapping chaotic, multi-dimensional data points onto the network's knots and lines, nodes and edges.

Obviously, this thought experiment quickly swells beyond the capacity of human visualization. Hence, this executive level review!

Probabilistic Inference

Conversely, the concept of *'probabilistic inference'* theorizes that the practical functionality of DL system derives from the cumulative distribution function described by its nonlinear activations (that is, the pattern of final values attributable to the network's nodes). This course discusses probabilistic inference in some detail in successive sections. For now, note that this school of thought led to the development of the 'dropout' method for NN regularization – a proven diagnostic point of order that weighs in favor of this theory's conjecture.

Operational Abstraction within Deep Learning

Generally speaking, the network layers within deep learning systems tend to transform input data into successively more abstract/compositional representations. In this case, as regards a deep architecture, these layers may in fact represent an individual multi-layered NN component of the deeper network – hence, the debate as to where learning occurs within these layers-of-layers.

For example, within image recognition, the input layer typically derives its matrix of from the image's pixels, reading its two-dimensional format as a one-dimension array of color values. The first representational DL layer may abstract pixels to encode edges (object edges, not edges of the graph, itself), while its second layer may compose and encode *arrangements* of edges (object corners or its horizontal/vertical aspects). Deeper still, a third layer may categorize the essential, unique structure of a nose or pair of eyes and, finally, a fourth layer may recognize a dog within the image, passing this diagnostic result as its final, user-facing output.

In this way, DL systems may *autonomously* learn practical data features (*i.e.*, the corners, edges, noses, eyes in an image) by optimally organizing these feature detectors/filters within the nodes of its layered components. Incidentally, to some degree of abstraction, contemporary neurological research hints that similar concepts may develop within the process of biological cognition, as well.

Artificial Neural Networks and Deep Neural Networks

Artificial Neural Networks

In review, an '*Artificial Neural Network*' (*ANN*) is simply a graphical model, or a collection of nodes connected to one another via edges. Each node may transmit information to another node along these edges, or not, depending upon design constraints – and this transmission may progress in a one-way or two-way fashion. These nodes have a '*state*', generally represented by a real number having a value between zero and one. Additionally, both nodes and edges may also manipulate a '*weight*' value, in addition to the node's state value (only nodes are '*stateful*'). These state and weight values provide targets for training, increasing or decreasing with each training iteration according to some preselected algorithm or protocol – these values also store the model's functional, learned arrangements in production mode. Additionally, ANNs may transmit data in a one-way or two-way fashion, according to design criteria.

Multi-layered ANNs (as opposed to simple ANNs, which possess only a single, fully-connected layer) typically transform the original input on a layer-by-layer basis. Within more complex ANN implementations, these layers may identify (on a more or less one-to-one basis) unique features (parameters) of interest hidden within the dataset. At a certain CAP (depth of layers), an ANN becomes sufficiently complex to transcend into the realm of deep neural networks, but no standard formally defines this point of inflection.

Deep Neural Networks

A '*Deep Neural Network*' (*DNN*) is an ANN with multiple human-inaccessible layers between the input and output layers – in other words, DNNs contain multiple '*hidden layers*' sandwiched between two '*visible layers*', one accepting input to the model and the other presenting model output. Though no standard delineates the architectural transition from '*shallow learning*' (ANNs) to '*deep learning*', most professionals require a deep design to manifest a CAP larger than two. Within each (typically) ANN component of its deep architecture, a deep learning system functions much like the simpler ANNs described above, excepting that these ANNs do not, themselves, necessarily present visible layers (which are reserved for user-/developer-level input/output). Unlike ANNs, these systems are typically '*feedforward*' (one-way data flow without recurrent looping).

Lesson Conclusion

Though neural network theory began to evolve almost immediately after advent of affordable commercial computing, researchers first coined the term 'deep learning' in 1986. After the application of GPUs to DL training in 2009, the Google Brain team released TensorFlow in 2015. This is a free, open source version of its proprietary deep learning software.

A deep learning system is a machine learning system implemented as a multilayer cascade of nonlinear processing units (graph models). Investigators typically use these models to perform feature extraction and transformation on large, complex, multivariate datasets that do not lend themselves well to 'traditional' application-specific solutions.

Deep learning networks are, essentially, multi-layer feed-forward artificial neural networks. Like ANNs, these adaptable and highly capable models benefit primarily from the autonomous discovery ('learning') of dataset features. This affordable process obviates the need to conduct expensive manual feature engineering.

Monte Carlo Methods

Monte Carlo Methods

Graph Models for Deep Learning

Lesson Overview

As previously stated, where systems tend to be simple and highly constrained, assumptive, parametric models suffice. If the data do not fit the assumptions of these application-specific models, however, then the applied method must somehow cope with the '*stochastic*' (random) nature of the underlying system. Recall previous descriptions of Bayesian systems, random forest- and tree-based methods, and k-Nearest-Neighbors algorithms, to list but a few. All these protocols strive to extend the application-specific model into chaotic, multivariate data spaces with varying degrees of success and applicability.

Finally, recall that these (more stochastic) methods led directly into related discussions of Support Vector Machines (the simplest form of machine learning construct) through graphical models and straight into neural networks. As a brief review of recent technological history reveals, hardware capacities finally rose to meet the rigorous demands of these complex implementations to pole-vault deep learning methods into suddenly ubiquitous and quite viable real-world solutions.

Overall, thus far, this course attempts to frame this steady progress from application-specific models to machine learning methods as a struggle to manage and analyze large, complex, multivariate and, most vitally, inherently chaotic systems. That being said, *chaos*, or *randomness*, is a common phenomenon that influences many things, both natural and artificial.

For example, as the student completes this course – which is, ideally, 'chock full' of information – the learning experience is a struggle to parse signal (educational information) from noise (nonsensical interference, like a blaring horn or stereo, and extraneous *data* – even extraneous *information*).

Relevant Nomenclature

This analogy begs a discussion of these terms, '*data*' and '*information*'. While synonymous in many contexts, within the realm of information technology, the term 'data' refers to raw, un-tabulated, un-curated, uncollated (*etc*...) input. By itself, data typically appears meaningless to human perception.

For example, an endless list of valid but randomly distributed *data* (perhaps numbers or names or dates, whatever) will provide the student with limited *information* (*e.g.*, the student will know the data is abundant and profuse, but these descriptors seldom assist with the resolution of problems related to the data). Information, then, is a distillation of data into something amenable to *human understanding*. In the previous analogy, a simple pie-chart drawn from a logically organized subset of the data might demonstrate a trend useful for explaining past performance or predicting future values. (Incidentally, the word 'data' is plural – the singular form is '*datum*' – thus, the informed student will say 'many data' or 'one datum'.)

Finally, within the context of the following discussion, a '*distribution*' represents the pattern that a given set of probabilities produce when demonstrated within a system – or as applied within a model. The most familiar example of this is the '*bell curve*' (or '*normal*') distribution, which, as implied by its label, resembles the silhouette of a bell.

To understand a function and its distribution, think of a baseball player and his or her performance curves. For example, a player will hit many balls during a given season to generate an individual batting average. Here, the player's batting method represents a function and the resultant pattern of hits represents the distribution of results, according to this individual skill.

Next, the student will review Monte Carlo Methods (MCM), which have a special place in both the advent of deep learning theory, itself, and, more vitally, in the optimized training routines that revolutionized the technology. Furthermore, in a certain sense, MCM bridged the gap between application-specific, rule-based methods and the true machine learning applications that provide the co-focus of this subject matter.

Overview (Monte Carlo Methods)

'*Monte Carlo Methods*' (*MCM*), as the label might imply, derive from the popular casino of the same name. The label originates in the Manhattan Project from one of its pivotal investigators, a scientist named Dr. Stanislaw Ulam (Los Alamos, 1946). While pondering the daunting task of modeling performance of a complex nuclear chain-reaction, Dr. Ulam pondered multiple possible solutions using differential equations, all of which quickly swelled to intractable proportions when considered from an exhaustive exploration of a vast problem space (*e.g.*, certain problems expand exponentially in complexity as they grow larger – like the game of telegraph, wherein one person tells two others, who then tell two others, and so on., *ad infinitum*). Ultimately, perhaps given the proximity of Los Alamos to Reno, Dr. Ulam conceptualized another method for exploring an inherently large and complex series of inter-connected events.

Simply put, he understood that he could 1) spend impossible quantities of time to derive a deterministic, 'hard-coded' formula/algorithm for describing these probabilities of interaction; OR 2) he could simply "game" the system (computers are very useful here) by running a simple, unrefined model repeatedly starting from randomly selected points, afterward performing a simple tabulation of the results to calculate resultant probabilities and further refine his model. While these randomized iterations of the model could not exhaustively explore the entire problem space, a sufficient number of *trials* (repetitions) efficiently defined model boundaries (or constraints or parameters – those useful features of the system).

Of course, no system is perfect, and this method is no exception. Given its stochastic nature, the results of such models are, also, stochastic – or somewhat fuzzy – in that each bit of information thus produced typically outputs with an error range, rather than as a single, definitive point value.

As a real-world example, consider the game of Canfield Solitaire. Theoretical upper limits postulate an optimal win rate of approximately 70-71%, while the best human players win less than 35% of their games. Clearly, humans are not optimized for this undertaking – and neither are estimations of the first model (whatever it was)!

NOTE: MCM typically depends on randomizations, as might be expected. Within computational systems, however, random is not always random. For example, the common *random number generator* available to most C++ compilers produces *pseudo-random* output, instead. This output depends upon a seed value – provided the seed is different at each initialization of the system, then the pattern of output approximates randomness. Unfortunately, inadvertent use of the *same* seed produces identical runs of output – this is decidedly *not* random performance.

Examples

'*Importance sampling*' presents a generalized technique for estimating a distribution's properties. Typically, these methods use samples drawn from a source other than the target source, another attempt to extrapolate from a more tractable system (that is less interesting) to a related, but more complex, system that *is* of interest.

'*Markov Chain Monte Carlo Methods*' (*MCMCM*) provide a stochastic (or randomized) model describing a sequence of possible events, wherein the probability of each event depends only on the state produced by the preceding event (hence the '*chain*' aspect of this label). These methods make possible the sampling of the *actual* system of interest (as referenced in the passage above), rather than a kinder, simpler substitute. Once derived via minor variation of Markov method described above, a '*Markov Chain*' (*MC*) will 'imitate' (model) the underlying distribution with its '*links*' (similar to a graph's 'nodes' and 'edges') – longer chains model the distribution with increasing fidelity, to a point. The investigator then draws samples from the distribution by simply retrieving a link-value after specific numbers of steps (trials) along the trained chain (model nodes). (Recall the fisherman's net analogy from previous sections of this course.)

'*Gibbs Sampling*' (*GS*) is commonly used in Bayesian inference systems (previously discussed) and, as such, provides another means of sampling complex systems that are otherwise intractable to application of deterministic (application-specific, rule-based) algorithms. This MCMCM algorithm provides a means for sampling a sequence of observations approximated from a given multivariate distribution. Again, these methods find utility where direct sampling is difficult or impossible, especially

Graph Models for Deep Learning

where the sample should represent a sequence of dependent events.

Finally, '*Contrastive Divergence*' (*CD*) is a form of '*gradient descent*' with '*line search*'. These are *iterative* (repetitively, serially executed) algorithms employing stochastic methods to identify 'minima' and 'maxima' within a distribution 'landscape'. CD proves particularly useful when inferring from high-dimensional (multivariate) distributions, given its Monte Carlo Method roots. Since the student will have cause to repeatedly interact with CD concepts, a brief analogy will be of immense benefit.

Previously, discussion referenced a '*distribution landscape*'. Visualize this by imagining a single, typical 'bell curve', or 'normal distribution'. This is, simply, a line shaped like the outline of a bell's silhouette. Now imagine a long column of such bells, each dented or damaged such that each presents a slightly variant rendition of the archetypal 'bell curve'. Stacking these lines together in two-dimensions along a steady interval will reveal a 'landscape' that resembles mountainous terrain rife with countless valleys and peaks, all randomly scattered across this rumpled plane.

Should the student have the need to identify the lowest points (*minima* or, singular, *minimum*) and highest points (*maxima* or, singular, *maximum*) in the landscape, a variety of methods for performing this service might leap to mind. If the number of measurements proved to be few, a manual approach might suffice. If the student must survey the surface of a planet the size of Jupiter, then she will want something much more efficient.

By now, the perceptive student will better understand the term '*gradient descent*', since the wayward surveyor must constantly descend (and, alternately, *ascend*) the three-dimensional *gradients* (or *slopes*) of the landscape while taking repeated measurements to tabulate its lows and highs. Naturally, a coherent hiking plan would prevent redundancies while ensuring consistent coverage of the terrain – hence the term 'line search', since the hiker might proceed in a line while performing these measurements. (The line need not be straight, however, since random exploration lies at the heart of this method – so the hiker will 'wander' across the landscape in a line, albeit one that is crooked).

Within this analogy, a repeatedly cloned hiker placed randomly within the terrain, if left to take measurements for some time, will eventually produce a list of elevations correlated with the problem space. By definition, each of these lists will have a lowest-of-the-lows and a highest-of-the-highs. These are all '*local minima*' and '*local maxima*', respectively. However, at the end of the survey, only *one* hiker will have crossed the *overall* (or global) high and low – and not necessarily the *same* hiker. These are the '*global minima*' and the terrain's '*global maxima*'. Of course, the size of a hiker's *steps* is important, too, since Paul Bunyan's survey would pass over features that the average student's stride would tumble. In this case, the cloning of the hikers and the individual hiker's steps both represent this method's *iterative* nature.

A final word on this analogy. If the survey enterprise possessed an infinite number of tiny hikers, an *exhaustive* measurement of the Jupiter-like planet's terrain might seem possible. In other words, the company might place one hiker to take one measurement at every possible point of the globe. This process *guarantees* discovery of the *true* global values. Nothing else will – although CD makes a best effort. This analogy, then, crudely attempts to describe the Contrastive

Divergence algorithm, which the attentive student will soon meet again.

Lesson Conclusion

Monte Carlo Methods may be applied to graphical models in a variety of ways – these methods also provide viable graphical solutions, too. Of principle importance to this subject matter, the Contrasting Divergence (CD) algorithm will become a familiar experience to most deep learning professionals, since this method and its variations exemplify current best practices as regards a vital aspect of deep learning (DL) implementations.

Specifically, the Markov methods described above, and their progenitors such as CD, appear repeatedly within the realm of DL training tasks, since these algorithms all attempt to optimize the exploration of a network-like construct. The perceptive student will anticipate how these protocols might impact subsequent discussions.

Approximate Inference and Expectation Maximization

Approximate Inference and Expectation Maximization

Lesson Overview

Although the concepts of *'Approximate Inference'* and *'Expectation Maximization'* (and many of the stochastic models described thus far) heavily rely upon *'Bayes' Theorem'*, the student need not be a Bayesian statistician to grasp the essential nature of these methods. Indeed, within this course, students should resist the urge to bog down in the theoretical detail underlying these constructs and, instead, focus on the subject matter. Namely, this course describes 'Graph Modeling' as applied within domains of 'Deep Learning' – thus, the current methods need be understood only to this level of implementation, for now.

During the following sub-sections, keep in mind the overall goal of these protocols, which attempt to make predictions (or classifications) of complex, stochastic (noisy) phenomena. These methods simply apply probabilistic thinking a bit 'deeper' into the same essential problem to better accommodate randomness and, eventually, use it to advantage.

Relevant Nomenclature

'Latent variables' are variables that cannot be directly observed. Rather, the investigator can only infer their presence and values from the measurements of *'observable variables'* – those that *can* be observed (quantified). Of course, both latent- and observable-variables may have unknown *values*, although the quantifiable nature of the latter kind supports construction of a (perhaps rudimentary) model to estimate their values based on samples drawn from these observations. Again, without exploring the underlying mathematical theory in detail, for now simply bear in mind that many stochastic methods attempt to solve for the hidden

variables by building a model using parameters constructed from observable variables.

Overview (Approximate Inference)

'*Approximate inference*' ('*variational inference*') supports a general class of stochastic (randomness-tolerant) graphical models commonly applied to 'big data' problems. While these solutions trade computational time for accuracy (decreasing the former due to computational complexities while increasing the latter), they often provide the only means of performing these essential tasks where exact inference (precise learning or '*automated feature engineering*') proves intractable.

Examples (Approximate Inference)

'*Markov random fields*' ('*Markov network*' or '*undirected graphical model*') represent a set of random variables having a Markov property described by an undirected graph. A '*Markov property*' is the memoryless property of a stochastic event or process (*i.e.*, an object property having a conditional probability distribution of future states that depend only upon the present state, not on the sequence of events that preceded it). In other words, if the thing is blue now, it is not blue now because it had previously been any other color (or, for that matter, because it will next be pink or yellow, *etc*...).

'*Markov networks*' are, essentially, *undirected* graphs that may be *cyclic*. These constructs differ from the Bayesian networks described below, which are *directed* and *acyclic*. (The astute student will note how only an undirected graph may be cyclic, but remember that many deep learning implementations contain both, so statements of this nature quickly become qualified.)

'*Loopy and generalized belief propagation*' ('*sum-product message passing*') is a message-passing method conducting inference using graphical models. This protocol is formulated as an *'exact inference'* (as opposed to the *'approximate inference'* described above) algorithm using tree or *'polytree'* structures, typically used to calculate marginal distributions of the graph's unobserved nodes, conditional on any observed nodes. A *'marginal distribution'* represents the probabilities of a parameter's value, independent of the value derived from other parameters within the system (its converse is a *'conditional distribution'*, which expresses the same value in a system of *dependent* parameters). Recall previous discussion of 'forest', 'tree' and 'nearest neighbor' methods.

'Variational Bayesian methods' provide tools for approximating intractable integrals arising in specific problems related to Bayesian inference and machine learning protocols. This enhanced alternative to Monte Carlo sampling methods (especially Markov Chain Monte Carlo Methods) supports identification of a locally-optimal, exact analytical solution to an approximation of the Bayesian posterior (essentially, a probability).

A *'Bayesian network'* (*'Bayes network'*, *'belief network'* or *'Bayesian model'*) is a probabilistic graphical model representing a set of variables and their *'conditional dependencies'* (the likelihood of the variable containing a specific value, given the values of its neighbors). The model is typically a *'Directed Acyclic Graph'* (*DAG*). Within these constructs, nodes represent Bayesian variables (observable quantities, latent variables, unknown parameters or hypotheses), each associated with a probability function. These node-level probability-functions input a particular set of values for the node's parent variables, returning as output the probability (or probability distribution) of the variable represented by the node. The overall model might accept, as input, a list of conditions (such as temperature and humidity) to return the probable outcome associated with those conditions (rain, perhaps).

Overview (Expectation Maximization)

'*Expectation Maximization*' (*EM*) is an iterative method for finding maximum likelihood (or '*maximum a posteriori*' [*MAP*]) estimates of parameters in statistical models, where the model depends on '*unobserved latent variables*'. This implementation of approximate inference finds utility within complex systems containing unknown parameters of interest (the latent variables) that are not otherwise amenable to statistical analyses. Again, a brief description of the iterative process will help clarify the fundamental purpose for applying EM within a deep-learning context.

Iteration

EM iteratively (repetitively, serially) performs an '*expectation step*' (*E step*) alternating with a '*maximization step*' (*M step*) until the probabilistic result of the model '*converges*' (its variable values approach their mode [central, or most likely, value] with decreasing uncertainty or 'spread'). Put another way, this system attempts to identify the *most likely* values for the 'latent variables' (hidden variables) of the system by constructing a model that also employs its 'observable variables'. Once the investigator provides an arbitrary (random) set of values to one set of variables (the latent variables, in this case), an iterative refinement process follows, wherein one set of function-output feeds to the next and then back again until the values of both sets converge to fixed points (usually according to some pre-defined cutoff), without actually deriving the definitive *actual* value (thus, the nature of a 'model', which is an imitation of a 'real-thing'). In fact, output that approaches this mode too closely may suffer from 'overfit'.

Deficiencies

Again, as with all models, deficiencies remain. The final set of variable values *may* represent a maximum or a '*saddle point*' (a 'flat spot' that stymies iterative progress without necessarily identifying the desire maxima – *i.e.*, the hiker finds a point in the terrain that lacks any nearby, mountable gradient). Additionally, as with the terrain analogy previously described, within the resultant landscape of 'all possible' variable values, multiple maxima may exist, yet a single given run of EM analysis is not guaranteed to identify the global maximum. Finally, due to the inherently stochastic nature of this procedure, nonsensical maxima may result.

Expectation Maximization *versus* Variational Bayesian Methods

Both Expectation Maximization (EM) and Variational Bayesian (VB) Methods are alternating, iterative procedures that successively converge on optimum parameter values. As such, these related but disparate protocols often confuse the less-informed student.

Both begin with formulas for probability densities and both require significant mathematical manipulation during development. The differences between these methods reveals an improving focus on derivation of *latent variables*, as opposed to parameters, which typically relate to a system's *observable variables*.

VB Differences

VB computes estimates of the actual posterior distribution of all variables, *both parameters and latent variables, together*. VB derives point estimates using the *mean* rather than mode of these values. The significance of parameter values is, thus, reduced, as compared to EM methods.

EM Differences

EM computes point estimates of the posterior distribution for those random variables identified as *parameters, only*. EM then estimates the *actual* posterior distributions of *latent variables*. Further, these computed point estimates are parameter *modes* (not definitive values, simply the most likely value) rather than means, and no other information is available. Thus, these methods improve resolution of a system's hidden variables. Use of 'mode', as opposed to 'mean', also moves these methods closer to accommodating stochastic (or semantic) concepts like 'context' and 'meaning'.

Lesson Conclusion

Increasingly complex Deep Learning (DL) constructs typically require increasingly robust methods for training (whether supervised or no). Computational efficiencies are vital.

Besides their obvious utility as modeling systems, in and of themselves, the methods reviewed within this section also impact the training process, as executed against the overall DL model, as well. Thus, these methods further provide a means for optimizing graph model training by reducing inference time while maximizing performance/efficacy of model refinement.

Deep Generative Models

Deep Generative Models

Lesson Overview

This section reviews two general classes of graphical models, the 'discriminative' and the 'generative'. While use of these general labels should not necessarily imply use of a specific model architecture, each class of models typically employs somewhat common design approaches to meet operational requirements. A particularly powerful class of DL implementations combine these two constructs together, such that the generative provides feedback to the descriptive, and *vice versa*, as each iteratively refines the other to achieve some common outcome.

Differences (Discriminative vs. Generative Models)

While reviewing the following sections, the student will benefit from a brief description of the basic differences between these model classes. Discriminative models generally optimize for classification problems, while generative models provide classification methods combined with the capacity for generating novel data in imitation of some 'real-world' process. Discriminative models *cannot* generate this kind of data, since classification is *not* a generative process.

Overview (Discriminative Models)

'*Discriminative models*' ('*conditional models*') describe the conditional likelihood of an event given a specific observation. Discriminatory power derives from quantifying dependence of the latent (unobserved, hidden) variables upon the values of observed (quantifiable) variables by modeling the latent variable's conditional probability distribution, given a specific observation. To clarify this concept, the student will benefit from a rough analogy.

Imagine a racetrack of specific length, which will host a variety of cars (and drivers) for a performance. The student observer may want to predict the average position of a given car at a given point in time after the starting gun sounds. Linear solutions may model the car's location according to a steady-state input (perhaps a straight line with a slope equal to the car's maximum rate of acceleration), but this is, at best, a crude approximation, since steady-state performance of variable machinery driven by variable drivers through variable weather across variable track conditions, and *ad infinitum*, will seldom produce a straight-line trend.

Clearly, a stochastic approach may produce more realistic predictions. A discriminative model will use one or more (or all) of these performance curves (variability distributions) to provide a likely location, considering all included sources of variation within the system. Thus, these models allow for predictions based on much more discriminative output at the student's time-interval of interest.

Of course, this crude analogy assumes fore-knowledge that supports creation of the performance curves in question.

Estimation of these curves is, then, precisely the goal of a discriminative model.

Examples (Discriminative Models)

As indicated in the previous sub-section, linear regression represents one of the simplest examples of a discriminative model, albeit an example with limited applicability to complex datasets. Many neural networks (previously described) also serve as robust discriminative models.

A bit more complex in design than a simple regression, the *'Support Vector Machine'* (*SVM*) (*'non-probabilistic binary linear classifier'*) is a graphical, supervised learning model. Used for classification and regression analysis, these models learn to represent examples as points in space with categories divided by a clear, wide gap. Training proceeds with labeled data (known values) until this division/classification boundary meets or exceeds some preset expectation. Once trained, new unlabeled data map into the same space, a process that supports prediction/classification of these new points (thus producing their true label, previously unknown). (Imagine the previous analogy of the fishing net, replacing the net with a single line tied periodically into knots, thus forming a kind of chain [or 'pearls on a string'] – otherwise, the 'data fitting' process of training is similar.)

'Maximum-Entropy Markov Models' (*MEMM*) or *'Conditional Markov Models'* (*CMM*) are discriminative models that extend standard maximum entropy classifiers. Though not strictly the case, the word *'entropy'*, here, essentially represents a *'multinomial logistic regression'* model, reviewed in the course introduction. These models assume unknown values (learning targets) relate to one another in Markov chains (also previously reviewed) - *i.e.*, data are not mutually

conditionally independent, rather each connects to (associates with, results from) another in a meaningful way. As with a Markov chain, predictions/classifications derive from retrieving values from the nodes of trained models.

Conditional Random Fields' (*CRF*) are '*discriminative undirected probabilistic graphical models*' used in pattern recognition, machine learning, structured prediction and a growing list of investigative endeavors. CRF considers context within the problem space in much the same way that MEMM models its parameters based on their dependencies (interrelationships). For example, within Natural Language Processing (NLP) deployments, linear chain CRF models predict sequences of labels for sequences of input – *i.e.*, context derives from the ability of these models to describe (label) one part of speech (or conversation, or text, *etc*...) according to its antecedent and descendant parts.

Overview (Generative Models)

'*Generative models*' describe the conditional likelihood of an observation given the occurrence of a specific event. Contrast this with the discriminative model and note both similarities and differences. In fact, in many ways, discriminative models and generative models mirror one another, at least as regards input/output (if not in underlying architecture and functionality). The astute student will likely already be asking questions about the outcome of combining these models together in some useful way.

Referring to the racetrack analogy described above, a reversed perspective will generate some interesting ideas. Rather than an observer modeling performance curves to predict car location at a given time-interval, the generative outcome is startlingly different. Imagine a system that could examine the student's observation of a car's position to determine which drivers drove which cars through what weather along which track. In other words, this is equivalent to using the winning placement to identify (or, better, *reproduce*) a specific race!

Examples (Generative Models)

'*Mixture models*' are probabilistic models that represent subpopulations within an overall population. These methods make statistical inferences about the properties of sub-populations, given only observations of the overall population, again without sub-population identity information.

'*Hidden Markov Models*' (*HMM*) analyze systems assumed to be Markov processes with unobserved (latent) states (hidden links among the known links of a chain, for example). These models correspond to the simplest dynamic Bayesian Network (previously described).

'*Probabilistic context-free grammars*' are graphical models that describe the structure of sentences and words in a natural language. Since many natural and artificial phenomenon operate on principles similar to those that guide the mechanics of language, these models frequently appear in non-language-related problem spaces. A '*context free grammar*' is a set of production rules that describe all possible strings in a given formal language. As with the application-specific methods previously defined, any attempt to produce an exhaustive list of such rules will likely transcend capacity – however, reverse analysis of a specific language can, at the least, potentially identify those rules in *practical use*.

'*Averaged one-dependence estimators*' provide a means for conducting probabilistic classification learning. These methods typically apply to the resolution of the attribute-independence problems of popular naive Bayes classifiers. Simply stated for the purposes of this lesson, this method often supports more accurate classifiers than naive Bayes (with modest increase to computation times).

'*Latent Dirichlet Allocation*' (*LDA*) is a generative statistical model primarily used within Natural Language Processing (NLP). Investigators use LDA to identify similarity among sets of observations via analysis of unobserved groups related to the observation set. Imagine a document composed in an arbitrary, unknown (though actual, rule-based) grammar (language), which contain topics (subjects, parameters) of interest to the student – LDA attempts to classify (identify, label) these hidden topics by assuming each of the document's words (observations) somehow relate to them. (Incidentally, the student will please pardon all the excessive parenthetical digressions – like this one –

Graph Models for Deep Learning

since the synonymous addenda are intended to bridge a wider gap of student interests and experiences.)

'Generative Adversarial Networks' (*GAN*) are unsupervised machine learning methods implemented as a system of two neural networks contesting with each other in a *'zero-sum'* game framework. Recall previously discussed potential for combining a discriminative model with a generative model – this is, essentially, a GAN. These solutions use the adversarial (opposing) relationship of this dual arrangement such that each network iteratively updates and refines the other in a cyclic, back-and-forth fashion. An 'autoencoder', discussed in following lessons, is one common example of a GAN.

'Restricted Boltzmann Machines' (RBM) are, as they imply, variations of Boltzmann machines. A *'Boltzmann Machine'* (*BM*) is a generative stochastic 'recurrent neural network'. In turn, a *'recurrent neural network'* is a graph model with an edge/node architecture that forms a directed graph along a sequence, thus supporting temporally dynamic behavior – in essence, recurrent NNs are *'stateful'*, having an internal short-term memory (*'Long Short-Term Memory'* [*LSTM*]) useful for analyzing a time-dependent sequence of observations. Finally, RBMs restrict the standard BM model in that graph nodes must form a bipartite graph, while both visible and hidden nodes may share symmetric edges – though no edges connect same-layer nodes. RBMs find common utility within deep learning implementations. Specifically, 'deep belief networks' (reviewed next) are comprised of 'stacked' RBMs, altogether fine-tuned using 'gradient descent' and 'back-propagation' – all reviewed within this course.

'Deep Belief Networks' (*DBN*) are generative graphical models in a Deep Neural Network configuration composed of multiple layers of latent variables (hidden layers) in a graph

implemented as an RBM or GAN (*autoencoder*). Within a DBN architecture, each sub-network's hidden layer serves as the visible layer for the next, thus supporting fast, layer-by-layer unsupervised training via application of *Contrastive Divergence* (*CD*) to each sub-network, starting from the lowest pair of layers (the lowest visible layer being a training set). Thus, via unsupervised training, DBNs learn to probabilistically reconstruct inputs during training. Afterward, these trained layers then act as feature-detectors. Additional supervised training will 'teach' the DBN to perform classification, as well.

Note, these implementations have a single training set and a single output set, each constantly refined by the other. Imagine a painting having a desired style training against another image having a desired structure – after training, the autoencoder can 'paint like an artist' onto any photograph of choice, turning your family photographs into priceless 'Renoirs'! (Google's DeepDream and DeepStyle currently offer public interactions with these models.)

'*Convolutional Boltzmann Machines*' (*CBM*) ('*Convolutional Deep Belief Networks*' [*CDBN*]) are a type of DNN comprised of stacked (layered) '*Convolutional Restricted Boltzmann Machines*' (*CRBM*). '*Convolution*', in turn, is a mathematical operation using two functions to produce a third function that describes how the shape of one modifies the other – this process appears often in computer vision applications, since the convoluting 'function' essentially analyzes an image in smaller 'frames' or 'chunks of input'. This convolutional aspect of the system allows CBMs to scale well to high-dimensional images, and also makes these models translation-invariant. CBMs (and most convolutional NN variants) use probabilistic max-pooling to reduce the dimensions in higher layers (a form of internal feature reduction). Training proceeds as with DBNs, layer-wise.

Also, like DBNs, additional training ('fine-tuning') using back-propagation optimizes the CBM for discrimination-oriented applications, as well. Additional training using CD optimizes the CBM for generative tasks. Incidentally, this review covers CNNs in more depth in following sections.

Finally, '*Auto-Regressive Networks*' (*ARN*) are graphical models used to describe certain time-varying processes. Output variables depend linearly on their own previous values and a stochastic term. Thus, these models are essentially stochastic difference equations. This type of modeling requires only limited knowledge about variable influences acting on system parameters of interest. Instead, primary input is typically a list of variables hypothesized to affect each other '*intertemporally*' (in a mutually-dependent time series). Imagine a model that predicts stock price based primarily on a time-coordinated list of its previous trades – no need to consider weather, politics, human emotions, *etc...*!

Lesson Conclusion

Both discriminative and generative models may perform classification tasks, while generative models are also capable of generating new data based on modeled features. As they become increasingly complex, these architectures also become increasingly more demanding of computational resources, in turn, requiring optimized training algorithms to produce viable 'real-world' solutions. The methods described above provide insight into these architectures. They also support efficient methods for training, testing and evaluating DL constructs, as exemplified by the basic operations of an autoencoder. Finally, Deep Generative Models are complex arrangements of simpler models (typically, RBMs and/or GANs) used as deep-learning (or deep-belief) systems.

Applications to Character Recognition, Natural Language Processing and Computer Vision

Applications to Character Recognition, Natural Language Processing and Computer Vision

Lesson Overview

As its title indicates, this closing section of the course reviews higher-level applications of previously described models. Specifically, many of these methods commonly serve as stand-alone or component models within software applied to the resolution of difficult problems associated with character recognition, natural language processing and computer vision. These fields consistently present complex, noisy input that long confounded the application-specific, rule-based protocols reviewed within the introduction. No wonder, then, that the flower of machine learning theory stretches its roots deeply into this solution domain. Accordingly, this 'capstone' review should orient the student within the contemporary technological landscape to empower superior insight into its imminent trends and demands.

Although research into these complex tasks continues unabated, all of them intricately entwined with one another – or not, depending upon design requirements – the following summary review attempts to tell a story of progress. This story builds from the success of rule-based methods (and rudimentary graphic models like Support Vector Machines), which provided an automated means of processing hand-written zip-codes long before Deep Learning emerged as a viable recourse. Passing through the surprisingly versatile field of natural language processing – which finds expression in scientific research ranged from biological laboratories through high-end data centers around the world – this story ends with a brief discussion of computer vision applications. This is, to be sure, a case of 'last but not least', since the future promises '*true artificial intelligence*', which can see, hear, touch and – perhaps – smell and taste the world around it!

Applications to Character Recognition, Natural Language Processing and Computer Vision

Overview (Character Recognition)

The process of '*character recognition*' (*CR*) is the algorithmic conversion of images of typed, handwritten or printed characters (letters, numbers, symbols, *etc*...) into machine-encoded text. While this problem may seem trivial when viewed from the perspective of modern understanding, without advanced graphical models and artificial intelligence, this analytical undertaking readily stumbles. Indeed, in many ways, given the need to transcribe millions of hand-printed (or cursive-style) addresses and zip-codes per day, the student might argue that CR research provided the soil from which the deep learning flower grew (to continue a crude metaphor).

In a perfect world, everyone would print with precision like a typewriter, thus making this problem much, much easier to resolve. Of course, any forensic hand-writing analyst will attest to the incredulity of this sentiment. The same is true with the process of language comprehension and artificial vision – the tasks that we, as humans, take for granted as simple are, in fact, the very problems that stump rule-based procedures. These are also the problems that deep learning models resolve with surprising alacrity.

MNIST Database

Since this minor subject matter fits nowhere else, this review begins with a brief word about the '*MNIST Database*' (Modified National Institute of Standards and Technology Database), mentioned here because it presents a best practice in the field. One now imitated many times as an increasing number of similarly large, curated, trainable datasets become available for public consumption.

The MNIST Database is a public repository of small images depicting a variety of handwritten digits. Investigators historically used this database (and continue to do so, to some extent) to train machine learning algorithms for character recognition. As of the time of this writing, the repository contains 60,000 training images and 10,000 testing images, while the updated *EMNIST Database* extends this collection to 240,000 training images and 40,000 testing images.

By now, as previously stated, several large databases make available images of various objects masked and labeled to support training of Deep Learning models. From a future perspective, wherein 'true artificial intelligence' has become a commonplace convenience, perhaps students will perceive these databases as the learning devices of an advanced nursery-school. As with all things intelligent, artificial cognition apparently begins with 'ABC', '123', 'I-hear-you' and 'you-see-me'.

Examples (Character Recognition)

Many software applications perform *'Optical Character Recognition'* (*OCR*), all using somewhat common, if uniquely implemented, variations or combinations of the models described thus far. In its most basic form, OCR performs analysis and conversion of *typewritten* text on a per-character basis. Extending this process, *'Optical Word Recognition'* (*OWR*) analyzes and converts typed text at the word-level, as its label implies, using 'whitespace' (non-printed background) as a delimiter.

Advancing the field, *'Intelligent Character Recognition'* (*ICR*) analyzes *handwriting/cursive* script on a per-character basis using machine learning techniques. Likewise, *'Intelligent Word Recognition'* (*IWR*) performs this serves at word-level.

Overview (Natural Language Processing)

'*Natural Language Processing*' (*NLP*) is a general term for the application of computational resources to the analysis of human language. As with Character Recognition, previously reviewed, and Computer Vision, reviewed next, artificial decomposition and/or comprehension of natural human language presents a noisy, multivariate problem that does not readily lend itself to rule-based approaches. Despite the fact that all human languages are rule-based constructs, their combinations, which include all possible variations and exceptions, quickly exceed astronomical proportions – infinity threatens. (Imagine compiling a list of *all possible* words, phrases and sentences formable by valid rules found within the English language.)

By now, this should sound familiar to the diligent student, since this is the same threshold that motivates the explosion of graphical model applications into so many contemporary endeavors. Within the context of NLP, common applications include speech recognition, natural language understanding, and natural language generation.

Traditionally, investigators enjoyed considerable success in the application of 'traditional', application-specific, rule-based methods to NLP problems – this review does not dismiss these profound innovations. Indeed, the application of complex IF-THEN-ELSE logic will readily mirror the essential rules of many human languages – to a point. Application of early graphical models of the size then amenable to contemporary computational capacities (*e.g.*, decision trees and forests) produced key advances within the field of NLP, as well.

Now that computational hardware advances to the limits of theoretical thinking, machine learning techniques quickly revolutionized the field, advancing its scope almost weekly. These systems owe their performance to their roots in the stochastic models reviewed within this course. Rather than attempt to circumvent or bypass the prodigious volume of both noise and information produced within the global enterprise of human communications (not to mention the 'communication' systems apparent among populations of animals, plants, ecosystems, genetic networks, traffic flow, cosmic phenomena, *etc*...), the deep learning models reviewed below 'embrace the horror' of chaos, using it to some advantage, even.

The following discussion describes three broad fields of NLP research. These include variations of syntactical-, semantic-, discourse- and speech-analyses. A broad range of deep learning models apply to these problems – indeed, Support Vector Machines and Random Forests also serve here – but these architectures typically return to variations on 'convolution' (reviewed in detail below) and 'recurrence' (reviewed above).

Essentially, these methods provide two non-exclusive benefits. Recurrence typically improves a model's temporal awareness where timing-to-sequence considerations matter. Convolution typically reduces feature space to provide a more efficient 'receptive field' arrangement (also reviewed below) while maintaining rotational-, translational- and scope-invariance. Recently, clever implementations of recurrent topologies yielded convolution-like results, however, and *vice versa*. Nevertheless, the informed student will keep these concepts in mind while completing this section of the course.

Relevant Nomenclature

The difference between the concepts of '*syntax*' and '*semantics*' is subtle but vital, especially within the context of the following discussion. First, either system may exist independently of the other – physical laws make no restrictions on the use of either syntax or semantics – indeed, human languages wildly vary their composition in this regard, though all adhere to *some* set of rules.

These formal rules, then, largely comprise the syntax of the system. Language may convey meaning purely using syntax – that is the order and arrangement of the information – but many students would probably bore quickly within such a world. Such a language might be more precise and efficient in many ways, but, then again, perhaps not, since the accurate description of certain phenomena (like language, itself) might require an infinite effort using only this mode of conveyance.

Semantics, as a concept, is a bit more nebulous and abstract. This is the essential *meaning* of the communication, that is 'meaning' in its most abstract sense. Think of this as the *emotion* of a language. For example, the word 'love' is an example of syntax, as a word construct properly formed of vowels and consonants in the correct order and structure; the same word expresses itself semantically with poetic beauty, as well, simply by definition. (That being said, an artful definition of semantics exceeds the scope of this work.)

Analysis of syntax may seem the more tractable problem. Indeed, demonstrations of super-computer-versus-human prowess clearly reveal how effectively artificial systems perform while conducting complex operations heavily weighted toward the syntactical. A game of chess, perhaps.

More obviously, these systems readily fail while performing even the simplest semantic task. Will the student's computer benefit from an affectionate hug? Perhaps not.

Finally, like CR and Computer Vision (reviewed next), NLP operations inevitably hover around the concept of '*context*', a term used frequently in this course, thus far, without formal definition. Within the realm of human communications, context is *everything*, as some say. Context is another way of describing the semantics of a communication, whether of a conversation, a textbook, a poem or even a picture (and much more). The identification and appropriate manipulation of contextual space represents another of those stochastic problem domains that provide fertile fields of opportunity for application of many deep learning methods. This is also, generally speaking, the domain of 'recurrence' and 'convolution' – both of which can provide this essential information, when properly deployed.

Applications to Character Recognition, Natural Language Processing and Computer Vision

Syntactical Analysis

'*Lemmatization*' is the algorithmic identification of a word's '*lemma*', based on its intended meaning. This field of investigation separates human language (specifically, its intellectual production during the speech process) into two phases. The first involves the mental management of a language's syntax and semantics – the 'what to say' – while the second is its '*phonology*', or the actual production of spoken words with meaning (the 'how to say', not just random sounds). Within the first stage of this model, lemma have meaning but no definitive utterance. Thus, in a sense, these models attempt to associate parts of speech with their '*archetypal*' (most essential) meaning, a process that, if successful, greatly simplifies many NLP considerations. Incidentally, this process contrasts with the similar process of 'stemming', reviewed below.

'*Morphological segmentation*' is an analytical method that analyzes words as individual '*morphemes*' to identify their '*morpheme class*'. A '*morpheme*' is the smallest grammatical unit in a language, but this is not conceptually the same as a word. For example, 'untestable' is a single word demonstrating three distinct morphemes: 'un', 'test' and 'able' (equivalent to the concepts 'not', 'examine', and 'ability').

As implied by its label, '*part-of-speech tagging*' analysis processes natural language to identify part-of-speech for each word in a sentence. While this may seem a trivial goal, these methods provide foundations for making much more complex predictions/classifications within the corpus of human communications. Likewise, '*parsing*', within the context of NLP, is the algorithmic production of the '*parse tree*' of a given sentence. This form of grammatical analysis comes in two general forms. '*Dependency Parsing*' analyzes for relationships between words, while '*Constituency Parsing*' produces a '*Parse Tree*' using a '*Probabilistic Context-Free Grammar*' (*PCFG*), previously reviewed. These 'trees' are, essentially, decision trees, also previously reviewed.

'*Sentence breaking*' ('*sentence boundary disambiguation*') is the algorithmic detection of sentence boundaries. Since this course presents in English, the student may mistake this field of endeavor for seeming somewhat trite. If so, consider non-idiomatic languages that do not use white-space or the 'period' symbol (or commas, *etc...*, for that matter) to *syntactically* denote such ideational boundaries.

Previously mentioned, '*stemming*' is the algorithmic reduction of inflected words to their word stem, base or root. This stem may not be the word's morphological root. Think search engines and synonyms, where the term 'synonym' expands a bit deeper into semantical space than its non-NLP definition. A word may have many synonyms, but many of these synonyms are context-dependent, since the same word in a different discussion might mean something else, entirely. Yet, the utility of stemming can be much subtler, too. For example, if the student searches the Internet using the words 'need a blue bus', a list of local bus-stops on 'blue-routes' might be more useful than a list of bus manufacturers producing buses that are painted blue. Or perhaps not. Context!

Finally, '*word segmentation*' is the analytical method for separating text into separate words. Again, this may initially seem trivial to English-speakers, but many languages do not use white space or punctuation to perform this operation *syntactically*. A common variation of this analysis, '*terminology extraction*' provides a means for automating the extraction of relevant '*terms*' within a body of text – in this case, terms are not simply words, rather they represent the key information or central ideas of a document, given its context. Context, again!

Semantic Analysis

After discussion of the previous section, the student might wonder why *syntactical* methods returned again and again to the concept of *context*. As previously stated, within a purely syntactical (rule-based) world, *context is everything*. Indeed, many investigators expend significant resources to distinguish the semantics of a discourse or conversation based purely on analysis of syntactical phenomena.

While the motivations for all this effort are probably many, the curious student might consider systems wherein the underlying language is utterly *unknown*. Simply by identifying a phenomenon as *'communication'*, the investigator assumes its contents somehow represent coherent, meaningful information (not simply a stream of noise). This, in turn, implies the presence of some system of rules, further implying some form of structure.

Though both prospects present difficult, perhaps intractable, problems, computational researchers may at least *hope* to resolve the syntactical nature of the underlying information by focusing on this structure, perhaps. How would the investigator begin to identify the unknown communique's *semantical* payload? The answer, perhaps: deep learning (or its descendants)!

From a 'traditional', rule-based, application-specific NLP perspective, the difference between syntax- and semantics-based analyses probably disappears, since the latter is so nebulous, even as a theoretical ideation. Perhaps this is why context figures so large in the outcome of these investigations. The same is perhaps true of semantical analysis, but Deep Learning systems are, in a sense, rather nebulous constructs, too. Where one stochastic phenomenon meets another, great things happen!

Loosely related, '*lexical semantics*' is the analysis of individual words and their contextual meaning, while '*machine translation*' performs algorithmic translation of text from one human language to another (or, in a more abstract sense, from one *schema* to another, where a schema represents some formal system of rules or syntax). Somewhat related, in that all three protocols attempt to prize a 'higher meaning' from chaotic linguistic systems, '*Named Entity Recognition*' (*NER*) analyzes text-streams to identify proper names from text – a decidedly semantic operation within a society so smitten by smart-phone texting.

Does the hybrid-text phrase, '[eye emoji] [heart emoji] bill', mean the sender 'likes a guy named bill', 'sees hearts, bill' (and is that a duck's bill, a hat bill or, again, a guy [or girl] named bill), or... madness! Further, what *is* an *emoji*, anyway?

'*Natural language generation*' provides methods for converting information and/or semantic intents into readable human language (imagine using a gaming remote to author the latest rousing espionage thriller). Conversely, '*natural language understanding*' is the process of converting textual '*chunks*' (streams of text that may or may not be syntactically/semantically related) into machine-readable logical representations (imagine converting this page-turning executive review into an interactive app – perhaps imbued with the ability to verbally answer questions and fabricate new passages from itself).

On that note, as they say, '*question answering*' is the process of automatically responding to human-language queries (questions). Although the student may even now be interacting with a personal digital assistant to play a favorite song, the same artificial intelligence will stumble when asked about the current 'mood' of a significant other. In part, this difficulty arises from an inability to *'recognize textual entailment'*, which presents another domain of investigation focused on automating the determination of logical entanglement within text fragments. For example, in the previous example (which used the search phrase '[eye emoji] [heart emoji] bill'), ambiguity largely results from this form of semantical failure. Similarly, methods developed within the field of *'relationship extraction'* provide a means for identifying relationships among named entities. The same search phrase (above) presents this difficulty, since the [eye emoji] language-part might represent the sender of the message, who might also be best-friends with a guy (or dog or goat or whatever) named 'Bill'. Or perhaps, the sender simply enjoys paying his or her way in life. (Bill! Get it? Probably, because the student is a master of both *syntax* and *semantics* – even when only *implied*!)

Finally, *'Sentiment analysis'* is the investigation and identification of public opinion (sentiment), as regards interesting trends within a textual corpus. How useful would *that* be? Likewise, *'topic segmentation and recognition'* and *'word sense disambiguation'* present two related domains of interest that provide the methods implied by their labels. Besides their application within Internet search engines, these implementations play vital roles within many forms of modern commercial (and academic) analyses focused on a variety of human phenomena – social media, for example.

Discourse Analysis and Speech Recognition

Given the wealth of knowledge imparted by this course, the student might imagine higher-level systems that combine two or more of these analytical concepts into a 'suite', 'toolkit' or 'workflow' intended to flexibly perform a variety of common tasks related to the interchange of information, whatever its nature. An industrious investigator might scour repositories of scholarly publications to automatically produce executive reviews like this one ('*automatic summarization*'). Then, he or she might automate the 'mentions' of one author and another to annotate their work ('*co-reference resolution*'). After publication, rave reviews might abound on forums and blogs scattered around the globe, so the proud investigator might build an amazing deep learning system to perform '*discourse analysis*' to identify fans or lucrative opportunities they might otherwise miss there.

Thus far, discussion focused on typed/handwritten text. While this input may derive from images (reviewed below), oftentimes this data flows from digital sources as existing streams of text. Closely related, but obviously more challenging, the field of '*speech recognition*' is a broad spectrum of endeavors focused on turning audible signals into digital (character-based) representations replete with meaning (both syntactical and semantic, when combined with the concepts previously described). Finally, '*speech segmentation*' and '*text-to-speech*' represent two of the more prolific sub-domain sources of innovation in this investigative realm.

Overview (Computer Vision)

This section reviews application of deep learning models to computer vision tasks with a specific focus on the utility, architecture and design motivations attributable to use of *'Convolutional Neural Networks'* (*CNN*). As implied by its label, the term *'Computer Vision'* describes a large, diverse and interdisciplinary field of investigation dedicated to the study and artificial reproduction of biological vision (and perception with actionable understanding).

The concepts of *'recurrence'* (previously reviewed) and *'convolution'* represent current best practices as regards machine learning implementations of computer vision techniques. As previously stated within the review of Natural Language Processing (NLP), these components provide two essential benefits within complex graphical models. Generally described, 'recurrence' provides a means of maintaining temporal (time-based) awareness within the dataset, while 'convolution' provides a means for reducing feature-space to maintain translational-, rotational- and scope-invariance during analysis. Current implementations of powerful, new deep learning models tend to combine these concepts with others to produce increasingly complex solutions that are, for all practical purposes, also increasingly more 'aware'.

Incidentally, the student may substitute the terms *'audio'* and *'auditory'* (*etc...*) for their visual equivalents to get the same review for the sound-side of the 'human experience'. This is the field of *'computer audition'*. The primary difference between the two senses is crucial, of course, since auditory input is, by nature, dynamic (temporal) and not static – *i.e.*, sound input is inherently multi-dimensional due to its waveform, which must include a time-based component.

As with NLP, computer vision applications abound. This section first describes three general domains of innovation in this regard to provide the context (that word again) for a subsequent review of convolutional design.

Recognition

'*Computer recognition*' ('*object classification*', '*object identification*' or '*object detection*') automates identification of a specific object, feature, or activity within graphical data. These systems, like most of the models described herein, have recently begun to outperform human experts in highly-semantical tasks such as photo-based identification of dog/cat breed. Wherever a human being currently labors at the operation of some device 'because only a human being can do it', these recognition methods present highly useful means of automation – they are, perhaps unfortunately, '*more human than human*' in some cases.

The general domain of '*content-based image retrieval*' is the algorithm search and retrieval of graphic data based on its visual content (*e.g.*, all pictures containing a dog). Related, '*pose estimation*', '*facial recognition*' and '*shape recognition*' technologies provide methods for automating the position/orientation of a specific object relative to the camera, recognition of faces for personal identification, and differentiation of people and objects within a scene, respectively.

Motion Analysis

Another closely related and highly active field of investigation extends these two-dimensional techniques into three- and four-dimensional space (the fourth dimension being 'time', as represented by videographic data). '*Motion analysis*' automates motion estimation via image sequence input – perhaps stills from a traditional film-based reel or frames captured from video output. Somewhat reversing this mode of analysis, '*egomotion*' instead automates the determination of rigid camera motion (rotation and translation) from an image sequence produced by the camera. Similarly, '*object tracking*' and '*optical flow*' study and predict, respectively, the algorithm movement of interesting entities through photographic/videographic space.

The eager student might put these concepts together – perhaps combining them with their equivalents within the auditory domain – to simply imagine possibilities. One day sooner rather than later, he or she might find employment running the classic Humphrey Bogart film, 'Casablanca', through a deep learning system capable of reproducing the set, action and characters (and their memorable faces/poses and voices), as well as camera angles and lighting, too. When re-projected using modern holographic displays, a lucky audience might sit at a table inside Rick's Café to watch and listen as Sam sings the forbidden 'As Time Goes By' for Lauren Bacall!

Media Recovery

On the subject of classical films, a brief note about '*media recovery*' ('*image recovery*', '*video recovery*' *etc...*) becomes appropriate. Many of these priceless bits of history eventually rotted within their film vaults due to the impermanent nature of celluloid, a popular source of film media during the 'Golden Age of Cinema'. Subsequently, many examples are 'lost to time' – traditional techniques of film restoration cannot recover a quality rendition of the originals. Here, deep-learning steps forward to shine. Besides providing a means for correcting image irregularities (increasing/decreasing or removing/adding effects such as motion-blur, red-eye, film-defects and more), these methods also automate '*scene reconstruction*', which is the algorithmic computing of a 3D model from an image/scene (something commonly accomplished via highspeed Lidar, which requires costly specialized hardware).

Deep learning and Computer Vision

Overview (Convolutional Neural Networks)

Reviewed in summary within previous sections of this course, '*Convolutional Neural Networks*' (*CNN*) currently dominate the field of computer vision (as well as the field of '*computer audition*', the artificial sense of hearing). As such, these constructs warrant final detailed description to ensure the student possess a firm grasp of its essentials, prior to completion of this course.

Visual (and auditory) information presents an obviously chaotic problem space. Like language, however, every rationally developed image (or video or sound recording) should contain an abundance of structure intermingled within the chaos – provided the scene is not simply a collection of randomized values produced as nonsense (and, even then, it will have structure by virtue of having been 'recorded').

For example, a video depicting the large crowd in attendance at a public event will seem chaotic at a glance, but it might present a thousand human faces, most having two eyes, one nose, one mouth, and so-on – all moving according to their own pathways, as well as effects due to motion of the camera. A rule-based method would search every arbitrarily-shaped sub-section of the imagery for all these features and more. Consequently, the traditional developer must explore countless unresolvable questions – should that arbitrary sub-section represent a square of 10x10 '*pixels*' (data points), 100x100 or should the space be a rectangle of 30x10, *ad infinitum*?

While the CNN architect must still make similar decisions at design-time, compromise currently ends there (perhaps, given more powerful computers, even *this* compromise might become irrelevant). Incidentally, this 'compromise' – the choice of sub-section size – is, in a generally descriptive sense, precisely the convolution described next.

Finally, a CNN is not one network (or graph model), but many, all working together within interconnected layers. Each layer has a specific design intended to perform a specific sub-service within the system.

Convolutional Layers

'*Convolutional layers*' apply a convolution operation to the input, passing the result to the next layer. As previously summarized, *convolution* emulates neural response to visual stimuli via application of a mathematical operation on two functions, which produces a third function that describes how one shape modifies the other.

Incidentally, for purposes of the following review, the student should not ponder the mathematical implications of the term '*function*', since the primary focus here is output of the layer – and this is simply an array (list, collection) of numbers. These numbers represent the function's output '*distribution*' (also previously reviewed), which is simply the range-based values that the function produces based on the layer's input. Recall the previous crude analogy to baseball, repeated below for convenience.

"To understand a function and its distribution, think of a baseball player and his or her performance curves. For example, a player will hit many balls during a given season to generate an individual batting average. Here, the player's batting method represents a function and the resultant pattern of hits represents the distribution of results, according to this individual skill."

Convolutional layers generally act to reduce graph complexity by reducing the number of free parameters within the data space. For example, image size becomes practically irrelevant (as compared to fully connected layers, reviewed below) by tiling (convoluting, scanning, moving) regions of size 5 x 5, each with the same shared weights (again, reviewed below). These iteratively combined convolutional regions represent a series of receptive fields and filters that, more or less, '*raster*' (pass across – sideways and down, in two-dimensional language) through the input image to scan for interesting, diagnostic features, which the CNN learns during training.

Pooling Layers

'*Pooling layers*' combine outputs of node clusters at one layer into a single node in the next layer. Two choices for pooling dominate within CNNs – '*max pooling*' forwards the maximum value from each cluster, while '*average pooling*' forwards the average cluster value, instead. Choice in this regard is not arbitrary, but depends upon the intended role of a given pooling layer, as well as overall design of the CNN.

Fully Connected Layers

Previously reviewed, '*fully connected layers*' connect every node in one layer to every node in another layer. This is, essentially, a '*multi-layer perceptron*', an early success among graph modeling protocols.

Receptive Field

The term '*receptive field*' has a specific meaning within discussions of neurobiology, and its meaning remains largely unchanged within this context – only the implementation differs significantly. Within CNNs, receptive fields are comprised of nodes in a previous layer that forward their output into a given set of nodes of the subordinate layer. This is not the same concept as 'pooling', which forwards the output of multiple nodes into one – rather, a receptive field presents a many-to-many connectivity (although the dimensions of the input-to-output node mapping typically reduces along the way – say from 10x10 to 5x5 – 'pooling' layers simply reduce output to a 1x1 format).

Weights

Recall the essential functionality of a graph model, which contains nodes connected by edges. Each node typically represents its value as a real number, which computes based on some internal, pre-defined function (most deep learning software make this function a matter of parametric choice). Additionally, each node may alter its raw output based on application of a '*weight*', which may increase/decrease the value accordingly. Typically, this weight is conceptualized as residing on the edge (whether of input or output, depending upon perspective). Taken together, this function-and-weight arrangement serve as a visual 'filter' (reviewed next).

For example, within the U.S. congress, each state elects two senators along with a number of representatives, which varies according to state-population to national-population ratios. Accordingly, simply in terms of a normalized distribution of power, senatorial votes typically 'count more' (are weighted more heavily) in the national schema, since most states send several representatives for every two senators (not to mention other differences, such as term length). Currently, the U.S. elects 100 senators and 435 representatives, a 100/435 ratio that roughly makes each representative vote equivocate to approximately 23% (0.23) of a senatorial vote. This is the representative-to-senator vote weight (note, this value naturally implies an inverse, that being the ratio 435/100, or 4.35, which is the *senatorial vote weight*). That's it for mathematics within this review, too!

Filters

Finally, a note about '*filters*'. These are not, strictly speaking, design-time components of the CNN. Rather, these visual (auditory) filters derive in an implied way from the relationship of node arrangements combined with their output/weight constructs. Filter size is closely related to the size of a selected '*convolutional field*' and they are often the same. CNN architectures generally share one filter among *all* its receptive fields, while fully connected layers use a separate vector of weights and bias for *each* receptive field.

The concept of filters lies at the heart of CNN feature reduction. Again, a crude analogy may be of benefit. Imagine a task requiring the student to sift the sand of a small beach for one lost earring (a feature of interest). The wealthy student might simply purchase a large number of graduated screen sets, each set containing a succession of finer mesh-densities, finally placing one set for each arbitrary space on the beach to sift the entire beach *at once*. Perhaps a better method might be to start at the top-left corner of the beach, sift that section of sand, and then take a step to the right and repeat this process until the task concludes. This process of choosing a location and area (the receptive field, filter size, convolution) combined with the iterative side-step (actually called a '*step*' within most deep learning software) is also the process of 'convolution'. Incidentally, this term describes both the convolutional field, and the convolutional function. The screens represent filters, their increasingly fine mesh density represents feature reduction provided by successive layers of analysis.

Perhaps conceptually, a filter is best imagined by recalling its 'real-world' equivalent. Within photography, these devices also act as 'feature reducers' by eliminating all blue light, for example, or fuzzing the image by passing its light emissions through a thin smear of glycerin. Further, most solar observatories use a simple, disc-shaped filter to blackout the sun's brilliant output, thus enabling the study of its corona and space surrounding the star.

Lesson Conclusion

This section completes an executive review of several highly innovative domains within the deep learning enterprise. Besides benefiting from the obvious factual information contained herein, this summary discussion of Character Recognition, Natural Language Processing and Computer Vision should improve the student's technological orientation while polishing and refining a deeper understanding of origins and directions. If 'true artificial intelligence' is the eventual outcome of contemporary interests, then the final product must combine all these models, or their descendants, into something unified and coherent.

One item of note, finally. Nothing herein describes application of deep learning models to the fourth and fifth senses. How should a silicon-chip 'taste' and 'smell'?

CODE BASE

Code Base

Overview

For means of quick reference and convenience, the following sub-sections provide live-code samples intended to execute against live, downloadable datasets. Perhaps, for most interested professionals, the 'big picture' presented below should seem... well, simple.

This is the creative power currently expressing itself via countless implementations of machine-learning solutions using a variety of deep learning software now available. The implementation – that is, the *syntactical* description and manipulation of even complex deep learning models should appear to be *easy*. In a manner of definition, this is certainly a true statement.

However, the development of practical deep learning solutions typically requires a bit more effort – most of it focused on manipulation of large datasets to make them presentable for ingestion and/or iterative refinement of model performance via repeated runs against multiple configurations and parameter settings (and more). This, then, might be the 'big picture' message presented below – software like Keras (and TensorFlow, Microsoft Cognitive Toolkit, *etc*...) make this process seem easy, although the final product will probably require greater application of art than science.

Keras Samples

Conveniently, *Keras* (at http://keras.io) includes a wide range of *Python*-based access to three popular deep learning software systems (*'back-end'* – the software that provides functional implementation of Keras-directed models), TensorFlow and Microsoft Cognitive Toolkit. Even better, to support novice developers, this well-managed product presents a variety of fast, easy-to-learn tutorials ready for copy/paste into the *Integrated Development Environment* (*IDE*) of your choice – simply compile and run them at your convenience. Even better than better, to support live examination of code (and underlying back-end software) functionality, Keras provides multiple real-world datasets, which the developer may access with trivial effort.

These datasets include the CIFAR10 (a repository of small images presented to assist development of classification models, specifically CNNs), IMDB (a repository of movie reviews to assist development of sentiment classifiers, specifically LSTMs), Reuters (a repository of newswires to support development of topic classifiers, specifically MLPs), MNIST (a repository of handwritten digits to support development of character classifiers, specifically MLPs & CNNs). And more – each sample begins with an overview that generally describes model objectives and design approaches, as well as indicating the particular data source in use.

Incidentally, the development process can be simply described, as well, although it can be, in practice, a rather complex undertaking. The code samples represent the construction, training and evaluation of the model – these are essential points of discussion marked appropriately within the code – with a final set of commands that report the model's final, evaluated performance. This is where development stops, at least within the context of this discussion. This is always where deployment begins.

At the end of the developmental process, Keras (via its underlying back-end software system) produces a trained 'model'. This is simply a file saved in the High-Density-File-5 (HDF5) format – this format, itself, presents a highly functional open-source API, which provides additional functionality to the highly-skilled developer. This model, then, may be installed and used within the production system as a stand-alone object (much like a word-processor application uses a variety of document file formats) via a number of methodologies. (Additionally, Keras supports output of JSON and YAML formats.) Again, a complete description of this development and deployment protocol exceeds the scope of this work.

Python

A word about Python. While a complete review of both Keras and the Python programming language exceeds the scope of this work, a few notes about essential concepts will be helpful. For those students familiar with programming methodologies, but unfamiliar with Python, the chief syntactical difference is Python's limited use of control characters – everything within these scripts is, for the most part, whitespace delimited. For example, indentation (the number of leading spaces or tabs) provides most code-block delimiting (except where the line ends in a 'comma' character, indicating a continuing list of parameters feeding a callable function – and even then, leading whitespace should be maintained as presented within the code below).

Finally, within most developmental languages, the comment character (in the case of Python, the '#' symbol) represents remarks provided by the developer to enhance coherence of a code-base. The following samples have internal comments from the originating demo, but they also contain remarks to support this discussion. For example, (### KERAS IMPORTS: BEGIN ###) indicates the start of code statements intended to set script's development environment. These comments are not required for model development, they merely support this text.

Code Base

Multilayer Perceptron (MLP) for Character Recognition of MNIST Database

This code sample trains against the MNIST Database previously reviewed. Configured as a Multi-Layer Perceptron (MLP), this Python-based Keras script begins (as all do) by calling library imports (*i.e.*, modules required to execute the attached code), here marked as (KERAS IMPORTS: BEGIN). Next, this code *initializes* (*i.e.*, sets initial values) and *regularizes* (*i.e.*, transforms the data for appropriate model input) the underlying dataset, here marked as (INITIALIZATION: BEGIN). Third, the script constructs its model(s), here marked as (MODEL CONSTRUCTION BEGIN), and finally it trains, evaluates and reports results of model development (MODEL TESTING, EVALUATION AND REPORTING).

Recall, Optical Character Recognition (OCR) is the algorithm processing of converting graphical renditions of letters and/or numerals into machine-interpretable equivalents – that is, the conversion of images of typed characters into logical (binary) representations of those characters. Also, recall that an MLP exemplifies construction of a basic "vanilla-flavored" neural network (or graph model).

Multilayer Perceptron (MLP)
for Character Recognition of MNIST Database

KERAS IMPORTS: BEGIN

from __future__ import print_function

import keras
from keras.datasets import mnist
from keras.models import Sequential

Code Base

Multilayer Perceptron (MLP) for Character Recognition of MNIST Database

```
from keras.layers import Dense, Dropout
from keras.optimizers import RMSprop

### KERAS IMPORTS: BEGIN ###

### INITIALIZATION: BEGIN ###

batch_size = 128
num_classes = 10
epochs = 20

# the data, split between train and test sets
(x_train, y_train), (x_test, y_test) = mnist.load_data()

x_train = x_train.reshape(60000, 784)
x_test = x_test.reshape(10000, 784)
x_train = x_train.astype('float32')
x_test = x_test.astype('float32')
x_train /= 255
x_test /= 255
print(x_train.shape[0], 'train samples')
print(x_test.shape[0], 'test samples')

# convert class vectors to binary class matrices
y_train = keras.utils.to_categorical(y_train, num_classes)
y_test = keras.utils.to_categorical(y_test, num_classes)


### INITIALIZATION: END ###

### MODEL CONSTRUCTION: BEGIN ###

model = Sequential()
model.add(Dense(512, activation='relu', input_shape=(784,)))
model.add(Dropout(0.2))
model.add(Dense(512, activation='relu'))
model.add(Dropout(0.2))
model.add(Dense(num_classes, activation='softmax'))

model.summary()

model.compile(loss='categorical_crossentropy',
          optimizer=RMSprop(),
```

Multilayer Perceptron (MLP)
for Character Recognition of MNIST Database

```
              metrics=['accuracy'])

### MODEL CONSTRUCTION:  END ###

### MODEL TESTING, EVALUATION AND REPORTING ###

history = model.fit(x_train, y_train,
            batch_size=batch_size,
            epochs=epochs,
            verbose=1,
            validation_data=(x_test, y_test))
score = model.evaluate(x_test, y_test, verbose=0)
print('Test loss:', score[0])
print('Test accuracy:', score[1])
```

Multi-Layer Perceptron for Topic Classification

This code sample trains against the Reuter's Newswire database to perform topic classification. Configured as a Multi-Layer Perceptron (MLP), this Python-based Keras script begins (as all do) by calling library imports (*i.e.*, modules required to execute the attached code), here marked as (KERAS IMPORTS: BEGIN). Next, this code *initializes* (*i.e.*, sets initial values) and *regularizes* (*i.e.*, transforms the data for appropriate model input) the underlying dataset, here marked as (INITIALIZATION: BEGIN). Third, the script constructs its model(s), here marked as (MODEL CONSTRUCTION BEGIN), and finally it trains, evaluates and reports results of model development (MODEL TESTING, EVALUATION AND REPORTING).

Recall that an MLP exemplifies construction of a basic "vanilla-flavored" neural network (or graph model). The previous code sample applied an MLP to an Optical Character Recognition problem. This script applies a similarly constructed MLP to a text-based classification task, which might provide a component within a discourse-analysis or cross-reference deployment.

Multi-Layer Perceptron for Topic Classification

```
### KERAS IMPORTS: BEGIN ###

from __future__ import print_function

import numpy as np
import keras
from keras.datasets import reuters
from keras.models import Sequential
from keras.layers import Dense, Dropout, Activation
from keras.preprocessing.text import Tokenizer
```

Code Base

Multi-Layer Perceptron for Topic Classification

```
### KERAS IMPORTS:  BEGIN ###

### INITIALIZATION:  BEGIN ###

max_words = 1000
batch_size = 32
epochs = 5

print('Loading data...')
(x_train, y_train), (x_test, y_test) =
reuters.load_data(num_words=max_words,
                                        test_split=0.2)
print(len(x_train), 'train sequences')
print(len(x_test), 'test sequences')

num_classes = np.max(y_train) + 1
print(num_classes, 'classes')

print('Vectorizing sequence data...')
tokenizer = Tokenizer(num_words=max_words)
x_train = tokenizer.sequences_to_matrix(x_train, mode='binary')
x_test = tokenizer.sequences_to_matrix(x_test, mode='binary')
print('x_train shape:', x_train.shape)
print('x_test shape:', x_test.shape)

print('Convert class vector to binary class matrix '
    '(for use with categorical_crossentropy)')
y_train = keras.utils.to_categorical(y_train, num_classes)
y_test = keras.utils.to_categorical(y_test, num_classes)
print('y_train shape:', y_train.shape)
print('y_test shape:', y_test.shape)

### INITIALIZATION:  END ###

### MODEL CONSTRUCTION:  BEGIN ###

print('Building model...')
model = Sequential()
model.add(Dense(512, input_shape=(max_words,)))
model.add(Activation('relu'))
model.add(Dropout(0.5))
model.add(Dense(num_classes))
model.add(Activation('softmax'))
```

Multi-Layer Perceptron for Topic Classification

```
model.compile(loss='categorical_crossentropy',
        optimizer='adam',
        metrics=['accuracy'])

### MODEL CONSTRUCTION: END ###

### MODEL TESTING, EVALUATION AND REPORTING ###

history = model.fit(x_train, y_train,
            batch_size=batch_size,
            epochs=epochs,
            verbose=1,
            validation_split=0.1)
score = model.evaluate(x_test, y_test,
            batch_size=batch_size, verbose=1)
print('Test score:', score[0])
print('Test accuracy:', score[1])
```

Code Base

Code Base

LSTM (Long Short-Term Memory) (Recurrent Neural Network [RNN])

Implemented as a *'Recurrent Neural Network'* (*RNN*) to demonstrate *'Long Short-Term Memory'* (*LSTM*) as applied to *'Sentiment Analysis'*, this Python-based Keras script begins (as all do) by calling library imports (*i.e.*, modules required to execute the attached code), here marked as (KERAS IMPORTS: BEGIN). Next, this code *initializes* (*i.e.*, sets initial values) and *regularizes* (*i.e.*, transforms the data for appropriate model input) the underlying dataset, here marked as (INITIALIZATION: BEGIN). Third, the script constructs its model(s), here marked as (MODEL CONSTRUCTION BEGIN), and finally it trains, evaluates and reports results of model development (MODEL TESTING, EVALUATION AND REPORTING).

Recall that *'Long Short-Term Memory'* (*LSTM*) represents the implementation of a *'Recurrent Neural Network'* (*RNN*), which provides *temporal* (or, as in this case, *sequential*) context to a model. Implemented against the *'IMDB Movie Database'* (*IMDB*), this script performs Sentiment Analysis by analyzing text-based movie review input for semantic relationships. In this case, recurrence represents context by providing sequential, word-part-based 'awareness' within the model.

Long Short-Term Memory (LSTM) Recurrent Neural Network (RNN)
KERAS IMPORTS: BEGIN ###   from __future__ import print_function   from keras.preprocessing import sequence   from keras.models import Sequential   from keras.layers import Dense, Embedding

Code Base

Long Short-Term Memory (LSTM)
Recurrent Neural Network (RNN)

```
from keras.layers import LSTM
from keras.datasets import imdb

### KERAS IMPORTS: BEGIN ###

### INITIALIZATION: BEGIN ###

max_features = 20000
# cut texts after this number of words (among top max_features most common words)
maxlen = 80
batch_size = 32

print('Loading data...')
(x_train, y_train), (x_test, y_test) = imdb.load_data(num_words=max_features)
print(len(x_train), 'train sequences')
print(len(x_test), 'test sequences')

print('Pad sequences (samples x time)')
x_train = sequence.pad_sequences(x_train, maxlen=maxlen)
x_test = sequence.pad_sequences(x_test, maxlen=maxlen)
print('x_train shape:', x_train.shape)
print('x_test shape:', x_test.shape)

print('Build model...')

### INITIALIZATION: END ###

### MODEL CONSTRUCTION: BEGIN ###

model = Sequential()
model.add(Embedding(max_features, 128))
model.add(LSTM(128, dropout=0.2, recurrent_dropout=0.2))
model.add(Dense(1, activation='sigmoid'))

# try using different optimizers and different optimizer configs
model.compile(loss='binary_crossentropy',
        optimizer='adam',
        metrics=['accuracy'])

### MODEL CONSTRUCTION: END ###
```

Long Short-Term Memory (LSTM) Recurrent Neural Network (RNN)

### MODEL TESTING, EVALUATION AND REPORTING ### 	
 	
print('Train...')
model.fit(x_train, y_train,
      batch_size=batch_size,
      epochs=15,
      validation_data=(x_test, y_test))
score, acc = model.evaluate(x_test, y_test,
                     batch_size=batch_size)
print('Test score:', score)
print('Test accuracy:', acc)

Code Base

Convolutional Neural Network (CNN) for Convolutional1D (Text Classification)

Implemented as a one-dimensional Convolutional Neural Network' (CNN) to perform text classification, this Python-based Keras script begins (as all do) by calling library imports (*i.e.*, modules required to execute the attached code), here marked as (KERAS IMPORTS: BEGIN). Next, this code *initializes* (*i.e.*, sets initial values) and *regularizes* (*i.e.*, transforms the data for appropriate model input) the underlying dataset, here marked as (INITIALIZATION: BEGIN). Third, the script constructs its model(s), here marked as (MODEL CONSTRUCTION BEGIN), and finally it trains, evaluates and reports results of model development (MODEL TESTING, EVALUATION AND REPORTING).

The previous sample used a '*Recurrent Neural Network*' (*RNN*) model to construct a '*Long Short-Term Memory*' (*LSTM*) network used to perform '*Sentiment Analysis*' against the '*IMDB Movie Database*' (*IMDB*). This script performs a similar analysis, that of '*Text Classification*', using a convolutional approach implemented as a simple 1D CNN.

Convolutional Neural Network (CNN) for Convolutional1D (Text Classification)
KERAS IMPORTS: BEGIN ### from __future__ import print_function from keras.preprocessing import sequence from keras.models import Sequential from keras.layers import Dense, Dropout, Activation from keras.layers import Embedding from keras.layers import Conv1D, GlobalMaxPooling1D from keras.datasets import imdb

Code Base

| Convolutional Neural Network (CNN) |
| for Convolutional1D (Text Classification) |

```
### KERAS IMPORTS:  BEGIN ###

### INITIALIZATION:  BEGIN ###

# set parameters:
max_features = 5000
maxlen = 400
batch_size = 32
embedding_dims = 50
filters = 250
kernel_size = 3
hidden_dims = 250
epochs = 2

print('Loading data...')
(x_train, y_train), (x_test, y_test) = imdb.load_data(num_words=max_features)
print(len(x_train), 'train sequences')
print(len(x_test), 'test sequences')

print('Pad sequences (samples x time)')
x_train = sequence.pad_sequences(x_train, maxlen=maxlen)
x_test = sequence.pad_sequences(x_test, maxlen=maxlen)
print('x_train shape:', x_train.shape)
print('x_test shape:', x_test.shape)

### INITIALIZATION:  END ###

### MODEL CONSTRUCTION:  BEGIN ###

print('Build model...')
model = Sequential()

# we start off with an efficient embedding layer which maps
# our vocab indices into embedding_dims dimensions
model.add(Embedding(max_features,
                    embedding_dims,
                    input_length=maxlen))
model.add(Dropout(0.2))

# we add a Convolution1D, which will learn filters
# word group filters of size filter_length:
```

Convolutional Neural Network (CNN) for Convolutional1D (Text Classification)

```
model.add(Conv1D(filters,
         kernel_size,
         padding='valid',
         activation='relu',
         strides=1))
# we use max pooling:
model.add(GlobalMaxPooling1D())

# We add a vanilla hidden layer:
model.add(Dense(hidden_dims))
model.add(Dropout(0.2))
model.add(Activation('relu'))

# We project onto a single unit output layer, and squash it with a sigmoid:
model.add(Dense(1))
model.add(Activation('sigmoid'))

model.compile(loss='binary_crossentropy',
         optimizer='adam',
         metrics=['accuracy'])

### MODEL CONSTRUCTION: END ###

### MODEL TESTING, EVALUATION AND REPORTING ###

model.fit(x_train, y_train,
       batch_size=batch_size,
       epochs=epochs,
       validation_data=(x_test, y_test))
```

Code Base

Convolutional Neural Network (CNN) for MNIST Database (Character Recognition)

Implemented as a *Convolutional Neural Network* (*CNN*) to perform '*Charcter Recognition*' (*CR*) against the *MNIST Database*, this Python-based Keras script begins (as all do) by calling library imports (*i.e.*, modules required to execute the attached code), here marked as (KERAS IMPORTS: BEGIN). Next, this code *initializes* (*i.e.*, sets initial values) and *regularizes* (*i.e.*, transforms the data for appropriate model input) the underlying dataset, here marked as (INITIALIZATION: BEGIN). Third, the script constructs its model(s), here marked as (MODEL CONSTRUCTION BEGIN), and finally it trains, evaluates and reports results of model development (MODEL TESTING, EVALUATION AND REPORTING).

A previous code sample performed this OCR task using a Multi-Layer Perceptron (MLP). Both perform the same analysis using vastly different approaches, however the MLP is more efficient for most such tasks. This example, then, demonstrates a basic CNN, as applied to a real-world problem.

Convolutional Neural Network (CNN) for MNIST Database (Character Recognition)

```
### KERAS IMPORTS: BEGIN ###

from __future__ import print_function
import keras
from keras.datasets import mnist
from keras.models import Sequential
from keras.layers import Dense, Dropout, Flatten
from keras.layers import Conv2D, MaxPooling2D
from keras import backend as K
```

Code Base

Convolutional Neural Network (CNN) for MNIST Database (Character Recognition)

```
### KERAS IMPORTS:  BEGIN ###

### INITIALIZATION:  BEGIN ###



batch_size = 128
num_classes = 10
epochs = 12

# input image dimensions
img_rows, img_cols = 28, 28

# the data, split between train and test sets
(x_train, y_train), (x_test, y_test) = mnist.load_data()

if K.image_data_format() == 'channels_first':
    x_train = x_train.reshape(x_train.shape[0], 1, img_rows, img_cols)
    x_test = x_test.reshape(x_test.shape[0], 1, img_rows, img_cols)
    input_shape = (1, img_rows, img_cols)
else:
    x_train = x_train.reshape(x_train.shape[0], img_rows, img_cols, 1)
    x_test = x_test.reshape(x_test.shape[0], img_rows, img_cols, 1)
    input_shape = (img_rows, img_cols, 1)

x_train = x_train.astype('float32')
x_test = x_test.astype('float32')
x_train /= 255
x_test /= 255
print('x_train shape:', x_train.shape)
print(x_train.shape[0], 'train samples')
print(x_test.shape[0], 'test samples')

# convert class vectors to binary class matrices
y_train = keras.utils.to_categorical(y_train, num_classes)
y_test = keras.utils.to_categorical(y_test, num_classes)

### INITIALIZATION:  END ###

### MODEL CONSTRUCTION:  BEGIN ###

model = Sequential()
model.add(Conv2D(32, kernel_size=(3, 3),
          activation='relu',
```

Convolutional Neural Network (CNN) for MNIST Database (Character Recognition)

```
            input_shape=input_shape))
model.add(Conv2D(64, (3, 3), activation='relu'))
model.add(MaxPooling2D(pool_size=(2, 2)))
model.add(Dropout(0.25))
model.add(Flatten())
model.add(Dense(128, activation='relu'))
model.add(Dropout(0.5))
model.add(Dense(num_classes, activation='softmax'))

model.compile(loss=keras.losses.categorical_crossentropy,
        optimizer=keras.optimizers.Adadelta(),
        metrics=['accuracy'])

### MODEL CONSTRUCTION: END ###

### MODEL TESTING, EVALUATION AND REPORTING ###

model.fit(x_train, y_train,
      batch_size=batch_size,
      epochs=epochs,
      verbose=1,
      validation_data=(x_test, y_test))
score = model.evaluate(x_test, y_test, verbose=0)
print('Test loss:', score[0])
print('Test accuracy:', score[1])
```

Code Base

Variational Auto-Encoder (VAE) a Generative Adversarial Network (GAN)

Implemented as '*Generative Adversarial Network*' (*GAN*) in the form of a modular, component-model '*Variational Auto-Encoder*' (*VAE*), this Python-based Keras script begins (as all do) by calling library imports (*i.e.*, modules required to execute the attached code), here marked as (KERAS IMPORTS: BEGIN). Next, this code *initializes* (*i.e.*, sets initial values) and *regularizes* (*i.e.*, transforms the data for appropriate model input) the underlying dataset, here marked as (INITIALIZATION: BEGIN). Third, the script constructs its model(s), here marked as (MODEL CONSTRUCTION BEGIN), and finally it trains, evaluates and reports results of model development (MODEL TESTING, EVALUATION AND REPORTING).

Recall, a primary benefit of a GAN solution is its ability to leverage its adversarial design to automate training of one network as applied against the other, and *vice versa*. These models have a single input/output set, which essentially represent the input/output graphic. The VAE analyzes its input to develop a '*style*' (*e.g.*, the basic brush-stroke style of Renoir or Picasso), and then apply this style to a target image – in other words, a properly trained VAE can turn your family photos into priceless classical masterpieces in a single pass using the discriminative-generative halves of its design.

Variational Auto-Encoder (VAE) a Generative Adversarial Network (GAN)
KERAS IMPORTS: BEGIN ###   from __future__ import absolute_import from __future__ import division from __future__ import print_function

Code Base

Variational Auto-Encoder (VAE)
a Generative Adversarial Network (GAN)

```python
from keras.layers import Lambda, Input, Dense
from keras.models import Model
from keras.datasets import mnist
from keras.losses import mse, binary_crossentropy
from keras.utils import plot_model
from keras import backend as K

import numpy as np
import matplotlib.pyplot as plt
import argparse
import os

# reparameterization trick
# instead of sampling from Q(z|X), sample eps = N(0,I)
# z = z_mean + sqrt(var)*eps
def sampling(args):
    """Reparameterization trick by sampling fr an isotropic unit Gaussian.
    # Arguments:
        args (tensor): mean and log of variance of Q(z|X)
    # Returns:
        z (tensor): sampled latent vector
    """

    z_mean, z_log_var = args
    batch = K.shape(z_mean)[0]
    dim = K.int_shape(z_mean)[1]
    # by default, random_normal has mean=0 and std=1.0
    epsilon = K.random_normal(shape=(batch, dim))
    return z_mean + K.exp(0.5 * z_log_var) * epsilon


def plot_results(models,
                 data,
                 batch_size=128,
                 model_name="vae_mnist"):
    """Plots labels and MNIST digits as function of 2-dim latent vector
    # Arguments:
        models (tuple): encoder and decoder models
        data (tuple): test data and label
        batch_size (int): prediction batch size
        model_name (string): which model is using this function
    """
```

Graph Models for Deep Learning 151

Variational Auto-Encoder (VAE)
a Generative Adversarial Network (GAN)

```
encoder, decoder = models
x_test, y_test = data
os.makedirs(model_name, exist_ok=True)

filename = os.path.join(model_name, "vae_mean.png")
# display a 2D plot of the digit classes in the latent space
z_mean, _, _ = encoder.predict(x_test,
                    batch_size=batch_size)
plt.figure(figsize=(12, 10))
plt.scatter(z_mean[:, 0], z_mean[:, 1], c=y_test)
plt.colorbar()
plt.xlabel("z[0]")
plt.ylabel("z[1]")
plt.savefig(filename)
plt.show()

filename = os.path.join(model_name, "digits_over_latent.png")
# display a 30x30 2D manifold of digits
n = 30
digit_size = 28
figure = np.zeros((digit_size * n, digit_size * n))
# linearly spaced coordinates corresponding to the 2D plot
# of digit classes in the latent space
grid_x = np.linspace(-4, 4, n)
grid_y = np.linspace(-4, 4, n)[::-1]

for i, yi in enumerate(grid_y):
    for j, xi in enumerate(grid_x):
        z_sample = np.array([[xi, yi]])
        x_decoded = decoder.predict(z_sample)
        digit = x_decoded[0].reshape(digit_size, digit_size)
        figure[i * digit_size: (i + 1) * digit_size,
            j * digit_size: (j + 1) * digit_size] = digit

plt.figure(figsize=(10, 10))
start_range = digit_size // 2
end_range = n * digit_size + start_range + 1
pixel_range = np.arange(start_range, end_range, digit_size)
sample_range_x = np.round(grid_x, 1)
sample_range_y = np.round(grid_y, 1)
plt.xticks(pixel_range, sample_range_x)
plt.yticks(pixel_range, sample_range_y)
```

Code Base

| Variational Auto-Encoder (VAE) |
| a Generative Adversarial Network (GAN) |

```python
    plt.xlabel("z[0]")
    plt.ylabel("z[1]")
    plt.imshow(figure, cmap='Greys_r')
    plt.savefig(filename)
    plt.show()


# MNIST dataset
(x_train, y_train), (x_test, y_test) = mnist.load_data()

image_size = x_train.shape[1]
original_dim = image_size * image_size
x_train = np.reshape(x_train, [-1, original_dim])
x_test = np.reshape(x_test, [-1, original_dim])
x_train = x_train.astype('float32') / 255
x_test = x_test.astype('float32') / 255

# network parameters
input_shape = (original_dim, )
intermediate_dim = 512
batch_size = 128
latent_dim = 2
epochs = 50




### INITIALIZATION:  END ###

### MODEL CONSTRUCTION:  BEGIN ###

# VAE model = encoder + decoder
# build encoder model
inputs = Input(shape=input_shape, name='encoder_input')
x = Dense(intermediate_dim, activation='relu')(inputs)
z_mean = Dense(latent_dim, name='z_mean')(x)
z_log_var = Dense(latent_dim, name='z_log_var')(x)

# use reparameterization trick to push the sampling out as input
# note that "output_shape" isn't necessary with the TensorFlow backend
```

Variational Auto-Encoder (VAE)
a Generative Adversarial Network (GAN)

```
z = Lambda(sampling, output_shape=(latent_dim,), name='z')([z_mean,
z_log_var])

# instantiate encoder model
encoder = Model(inputs, [z_mean, z_log_var, z], name='encoder')
encoder.summary()
plot_model(encoder, to_file='vae_mlp_encoder.png', show_shapes=True)

# build decoder model
latent_inputs = Input(shape=(latent_dim,), name='z_sampling')
x = Dense(intermediate_dim, activation='relu')(latent_inputs)
outputs = Dense(original_dim, activation='sigmoid')(x)

# instantiate decoder model
decoder = Model(latent_inputs, outputs, name='decoder')
decoder.summary()
plot_model(decoder, to_file='vae_mlp_decoder.png', show_shapes=True)

# instantiate VAE model
outputs = decoder(encoder(inputs)[2])
vae = Model(inputs, outputs, name='vae_mlp')

if __name__ == '__main__':
   parser = argparse.ArgumentParser()
   help_ = "Load h5 model trained weights"
   parser.add_argument("-w", "--weights", help=help_)
   help_ = "Use mse loss instead of binary cross entropy (default)"
   parser.add_argument("-m",
               "--mse",
               help=help_, action='store_true')
   args = parser.parse_args()
   models = (encoder, decoder)
   data = (x_test, y_test)

   # VAE loss = mse_loss or xent_loss + kl_loss
   if args.mse:
      reconstruction_loss = mse(inputs, outputs)
   else:
      reconstruction_loss = binary_crossentropy(inputs,
                              outputs)

   reconstruction_loss *= original_dim
   kl_loss = 1 + z_log_var - K.square(z_mean) - K.exp(z_log_var)
```

Code Base

Variational Auto-Encoder (VAE)
a Generative Adversarial Network (GAN)

```
    kl_loss = K.sum(kl_loss, axis=-1)
    kl_loss *= -0.5
    vae_loss = K.mean(reconstruction_loss + kl_loss)
    vae.add_loss(vae_loss)
    vae.compile(optimizer='adam')
    vae.summary()
    plot_model(vae,
            to_file='vae_mlp.png',
            show_shapes=True)

### MODEL CONSTRUCTION:  END ###

### MODEL TESTING, EVALUATION AND REPORTING ###

    if args.weights:
        vae.load_weights(args.weights)
    else:
        # train the autoencoder
        vae.fit(x_train,
            epochs=epochs,
            batch_size=batch_size,
            validation_data=(x_test, None))
        vae.save_weights('vae_mlp_mnist.h5')

    plot_results(models,
            data,
            batch_size=batch_size,
            model_name="vae_mlp")
```

Course Conclusion

Course Conclusion

A Closing Analogy

Since the belabored student has applied so much time and interest in completing this course, its conclusion will not regurgitate all that has been. Instead, it ends with yet another crude analogy.

A bit like a neighborhood scavenger hunt, the deep learning enterprise is wide open and loaded with goodies! Easy scores abound!

The successfully outbound student now stands at the ribbon with his or her knowledge-basket (or pocket book) in hand, ready to compete with countless others, all eager to bag the low-hanging fruit, as they say. Two attributes (features!) will essentially determine individual fortunes – experience and education.

Ideally, the student's embrace of this executive review will generate immediate rewards that lead to long-lasting benefits. In terms of the course, the student now possesses a revitalized 'discriminative-generative cognitive model', geared for analytical success. Think of its intended benefits as a bigger basket, better shoes, faster legs and a quicker eye!

Now, then... on your marks! Get ready! Go!

Course Conclusion

Glossary of Terms

adjacency 35, 37, 40
Adjacency lists 36
adjacency matrices 37
adjacent 35
Ancestral graphs..............42
ANCOVA 16
ANN 23, 57, 58
ANNs 57, 58, 59
ANOVA 16
Application-specific 16
*approximate inference*78, 80
Approximate inference.....77
Approximate Inference73, 75, 77
archetypal 69, 106
ARN 95
artificial neural network 57
Artificial Neural Network ..23
artificial synapse 23
audio 30, 113
auditory30, 113, 115, 117, 121
automated feature engineering.....................77
automated feature extraction......................28
automatic summarization 112
Auto-Regressive Networks 95
average pooling 119
Averaged one-dependence estimators.....................92
back-end 4, 127
back-propagation 93, 95
Bayes network 78
Bayes' theorem................22

Bayes' Theorem 75
Bayesian model 78
Bayesian networks42, 77, 78
belief network 78
belief networks31, 41, 42, 93
bell curve 13, 64, 69
Big Data Research and Development Initiative .48
boosted tree.....................22
CAP 51, 57, 58
cascade 50, 59
cascades 28
CBMs 94
CD 69, 70, 72, 94, 95
Central Processing Unit...48
CGA.................................47
chain ..41, 51, 66, 68, 90, 91
Chain graphs 41
chaos 63, 103, 117
character recognition4, 99, 100, 101
chunks 94, 110
Clique trees 41
CMM 89
CNN113, 117, 118, 119, 121
CNNs 95, 119, 120
Color Graphics Adapters .47
computer audition ..113, 117
Computer recognition 114
Computer Vision97, 102, 105, 113, 117, 123
conditional dependencies 42, 78
conditional distribution 78

Conditional Markov Models 89
conditional models 88
conditional random field...42
Conditional Random Fields 90
Constituency Parsing.....107
content-based image retrieval......114
context free grammar 92
Contrastive Divergence 69, 71, 94
convolution 103, 105, 113, 118, 121
Convolution.............94, 103
Convolutional Boltzmann Machines 94
Convolutional Deep Belief Networks 94
Convolutional layers 118, 119
Convolutional Neural Networks 113, 117
Convolutional Restricted Boltzmann Machines...94
co-reference resolution..112
CPU 48
CR 100, 105
CRBM 94
Credit Assignment Path...51
CRF 90
curve-fitting 17
DAG 42, 78
data fitting 22
DBN 93
DBNs 94
Decision tree....................19
Deep Belief Networks 51, 93
deep learning 4, 21, 29, 30, 32, 58, 59, 63, 72, 103, 113
deep neural network ..51, 58
Deep Neural Network 47, 93

Dependency Parsing 106
dependent variable 9, 17, 19
descriptive statistics...11, 12
Digital Visual Interface.....47
directed acyclic graph......78
directed acyclic graphs 41, 42
directed cycle..................41
directed graph......25, 40, 93
directed graphical models 42
directedness 25
discourse analysis 112
Discriminative models 87, 88
discriminative undirected probabilistic graphical models..........................90
DistBelief 48
distribution 13, 55, 64, 68, 69, 77, 78, 82, 88, 118, 120
distribution landscape......69
DNN 47, 58, 94
DNNs 48, 58
DVI..................................47
E step 80
edge 25, 35, 36, 37, 38, 39, 93, 120
edge collection................37
egomotion 115
EMNIST Database.........101
emoji 110, 111
ensemble learning 22
exact inference 77, 78
exhaustive analysis 18
Expectation Maximization 73, 75, 80, 82
expectation step 80
explanatory variable ..10, 18
facial recognition............114
factor...............................41
Factor graphs 41

feature 22, 28, 50, 56, 59, 94, 103, 113, 114, 121, 122
feature engineering 28, 59
features 11, 28, 52, 56, 57, 59, 66, 70, 96, 117, 119, 157
filters 56, 119, 121
F-test 17
fully connected layers 119, 121
Fully parametric models ..13
function 9, 21, 39, 41, 55, 64, 78, 94, 118, 120, 121
GAN 93, 94
generalized additive models 21
Generalized Linear Models 21
Generative Adversarial Networks 93
Generative models 91
Gibbs Sampling 68
global minima 18, 70
GPU 48
gradient descent 69, 93
Graphical Processing Units 48
GS 68
hard-coded 14, 16, 28, 32, 66
Hash table 39
HDMI 47
Hidden layers 51
Hidden Markov Models 42, 91
High Definition Multimedia Interface 47
HMM 91
ICR 101
IDE................................128
image recovery 115
IMDB.............128, 137, 141

IMDB Movie Database 137, 141
Importance sampling 68
in silico 35, 36
incident 35, 36, 37
incident lists 36, 37
incident matrices..............37
incidental 35
independent variable 9
Indexed arrays 39
initializes 130, 133, 137, 141, 145, 149
Integrated Development Environment 128
Intelligent Character Recognition 101
Intelligent Word Recognition 101
intertemporally 95
iterative 69, 70, 80, 81, 82, 121
IWR................................101
join trees 41
junction trees 41
Keras 3, 4, 127
k-nearest neighbor...........23
k-NN 23
labeled training data 29
Latent Dirichlet Allocation 92
latent variables 51, 78, 80, 82, 93
Latent variables 75
lazy learning 23
LDA.................................92
learning 3, 4, 5, 15, 19, 20, 21, 22, 23, 25, 26, 27, 28, 29, 30, 31, 32, 40, 42, 47, 48, 49, 50, 51, 53, 55, 59, 63, 65, 72, 77, 78, 80, 89, 90, 92, 93, 96, 99, 100, 101, 102, 103, 105, 109, 112, 113, 115, 117, 120, 121, 123, 157, 167

least squares methods 22
lemma............................106
Lemmatization 106
lexical semantics 110
linear model 16
linear regression13, 16, 17, 19, 21, 89
linguistic analysis.............25
List structures 36
lists-of-lists 36, 39
lists-of-matrices 39
local minima.....................18
logistic regression......19, 21
Long Short-Term Memory 93, 137, 141
Loopy and generalized belief propagation........77
LSTM 93, 137, 138, 141
M step...............................80
MANCOVA 16
MANOVA 16
marginal distribution 78
Markov chain 90
Markov Chain 68, 78
Markov Chain Monte Carlo Methods.......................68
Markov chains 89
Markov network 77
Markov networks 41, 77
Markov property...............77
Markov random fields 77
matrix structures ..37, 39, 43
max pooling 119
maxima 18, 69, 70, 81
maximization step............80
Maximum-Entropy Markov Models........................89
MC 68
MCM 65, 66, 67
MCMCM 68
media recovery 115
MEMM 89, 90

Microsoft Cognitive Toolkit3, 4, 40, 127
minima 18, 69, 70
mixed models 42
Mixture models 91
MNIST Database ...100, 101
Monte Carlo Methods61, 65, 66, 72, 78
more human than human 114
morpheme 106
morpheme class 106
morphemes....................106
Morphological segmentation 106
Motion analysis..............114
multi-layer perceptron 119
multinomial logistic regression....................89
multivariate linear regression....................17
Naive Bayes 22
Named Entity Recognition 110
Natural language generation 110
Natural Language Processing42, 90, 92, 97, 102, 113, 123
natural language understanding....102, 110
NER 110
neurons................23, 25, 35
NLP30, 42, 90, 92, 102, 103, 105, 106, 107, 109, 113, 114
node23, 35, 36, 37, 38, 39, 41, 53, 57, 78, 93, 119, 120, 121
node collection.................37
node/edge set..................37
non-linear..................10, 17
Non-parametric models ...14

non-probabilistic binary linear classifier............89
non-scalar variable..........10
normal...................13, 64, 69
numerical optimization.....18
object classification........114
object detection..............114
object identification........114
object tracking................115
Object-oriented list...........40
observable variables 75, 80, 82
OCR...............................101
Optical Character Recognition................101
optical flow.....................115
Optical Word Recognition 101
ordinal logistic regression 19
Ordinary Least Squares...22
Overfit...............................22
OWR...............................101
parameter..............9, *78, 82*
parsing......................26, 106
Part-of-speech tagging ..106
PCFG..............................107
phonology......................106
polytree............................78
Pooling layers................119
pose estimation..............114
predictive analytics..........20
Predictive modeling.........20
probabilistic classifiers.....22
Probabilistic Context-Free Grammar....................107
Probabilistic context-free grammars.....................92
probabilistic inference 41, 55
pseudo-random................67
Python...........3, 4, 127, 129
question answering........110
Random forest.................22
Random Forests............103

random number generator 67
randomness 13, 17, 21, 63, 67, 75, 77
RBM..........................93, 94
RBMs........................93, 96
receptive field 103, 119, 121
recognize textual entailment 111
recurrence.....103, 105, 113
recurrent neural network..93
Recurrent Neural Network 137, 141
regularizes 130, 133, 137, 141, 145, 149
relationship extraction....111
response variable.........9, 18
Restricted Boltzmann Machines.....................93
RNN........................137, 141
robust regression.............21
rule-based 3, 4, 14, 16, 20, 27, 28, 32, 65, 68, 92, 99, 100, 102, 109, 117
saddle point......................81
scalar variables................10
scene reconstruction.....116
Semantic Analysis.........109
semantics 104, 105, 106, 109, 111
Semi-parametric models..14
semiparametric regression 21
sentence boundary disambiguation...........107
Sentence breaking.........107
Sentiment analysis........111
Sentiment Analysis 137, 141
sequential......................137
shallow learning...............58
shape recognition..........114
speech recognition 30, 43, 102, 112

speech segmentation 112
stateful 57, 93
Statistical inference 11
statistical model ...11, 13, 92
stemming 106, 107
step 121
stochastic 22, 63, 66, 68, 69, 75, 77, 81, 83, 88, 93, 95, 103, 105, 109
style 94, 100, 149
sum-product message passing 77
Super-Video Graphics Adapter........................47
supervised 23, 29, 50, 84, 89, 94
Support Vector Machines 23, 32, 63, 99, 103
SVGA..............................47
SVM 23, 89
synapses 25, 35
Syntactical Analysis 106
syntax 104, 106, 109, 110, 111
TensorFlow 3, 4, 40, 48, 59, 127
terminology extraction ...107
terms 9, 10, 20, 28, 50, 64, 107, 113, 120, 157
Text Classification 141
text-to-speech 112
topic segmentation and recognition..................111
training 22, 23, 28, 29, 48, 50, 54, 57, 59, 65, 72, 84, 94, 95, 96, 101, 119

treewidth 41
true artificial intelligence 99, 101, 123
T-test 17
undirected 25, 35, 41, 42, 77
undirected graph 25, 77
undirected graphical model 77
universal approximation theorem 53
unobserved latent variables 80
unordered lists 36
unsupervised .29, 50, 93, 94
VAE 149, 152, 153
Variable-Order Markov Models 42
Variational Auto-Encoder 149
Variational Bayesian methods.......................78
variational inference 77
vertices 25
VGA 47
Video Graphics Adapters.47
Video In Video Out 47
video recovery 115
Visible layers 51
VIVO 47
weight 57, 120, 121
word segmentation 107
word sense disambiguation 111

END

ABOUT THE AUTHOR

Born in Texas and currently residing in Washington D.C., Stephen Donald Huff is an author of fiction novels, short stories and poetry. He is also a published scientist with expertise in bioinformatics (computational biology) and machine learning. Message him at Stephen@StephenHuff.com.

www.ingramcontent.com/pod-product-compliance
Lightning Source LLC
Chambersburg PA
CBHW031629210526
45464CB00004B/1815